T0341195

Remote Sensing Applications for the Urban Environment

Taylor & Francis Series in
Remote Sensing Applications

Series Editor
Qihao Weng

Indiana State University
Terre Haute, Indiana, U.S.A.

Remote Sensing Applications for the Urban Environment, *George Z. Xian*

Remote Sensing of Impervious Surfaces in Tropical and Subtropical Areas, *Hui Lin, Yuanzhi Zhang, and Qihao Weng*

Global Urban Monitoring and Assessment through Earth Observation, *edited by Qihao Weng*

Remote Sensing of Natural Resources, *edited by Guangxing Wang and Qihao Weng*

Remote Sensing of Land Use and Land Cover: Principles and Applications, *Chandra P. Giri*

Remote Sensing of Protected Lands, *edited by Yeqiao Wang*

Advances in Environmental Remote Sensing: Sensors, Algorithms, and Applications, *edited by Qihao Weng*

Remote Sensing of Coastal Environments, *edited by Qihao Weng*

Remote Sensing of Global Croplands for Food Security, *edited by Prasad S. Thenkabail, John G. Lyon, Hugh Turral, and Chandashekhar M. Biradar*

Global Mapping of Human Settlement: Experiences, Data Sets, and Prospects, *edited by Paolo Gamba and Martin Herold*

Hyperspectral Remote Sensing: Principles and Applications, *Marcus Borengasser, William S. Hungate, and Russell Watkins*

Remote Sensing of Impervious Surfaces, *edited by Qihao Weng*

Multispectral Image Analysis Using the Object-Oriented Paradigm, *Kumar Navulur*

Remote Sensing Applications for the Urban Environment

George Z. Xian

CRC Press
Taylor & Francis Group
Boca Raton London New York

CRC Press is an imprint of the
Taylor & Francis Group, an **informa** business

CRC Press
Taylor & Francis Group
6000 Broken Sound Parkway NW, Suite 300
Boca Raton, FL 33487-2742

© 2016 by Taylor & Francis Group, LLC
CRC Press is an imprint of Taylor & Francis Group, an Informa business

No claim to original U.S. Government works

ISBN 13: 978-1-4200-8984-4 (hbk)

Visit the Taylor & Francis Web site at
http://www.taylorandfrancis.com

and the CRC Press Web site at
http://www.crcpress.com

Contents

Foreword ix
Preface..xi
Author...xiii

1. **Introduction** 1

2. **Characteristics of Urban Landscape Feature in Local and**
   Regional Scales 7
   2.1   Introduction 7
   2.2   Role of Remote Sensing for the Urban Landscape...................7
   2.3   Application of High-Resolution Satellite Imagery 9
   2.4   Mapping Urban Land Cover Using QuickBird Imagery 12
   2.5   Decision Tree Model for Urban Land Cover Classification 14
   2.6   Object-Based Image Analysis for Urban Land Cover
          Classification..16
   2.7   Summary...22

3. **Characterization of Urban Land Cover in a Moderate Resolution**.....23
   3.1   Introduction 23
   3.2   Characteristics of Urban Landscape 23
   3.3   Mixing Features of Urban Landscapes....................................28
   3.4   Characterization of Urban Landscape 29
          3.4.1   Spectral Mixture Analysis...29
          3.4.2   Regression Tree Algorithms......................................34
   3.5   Another Method for Urban Land Classification.....................46
          3.5.1   Artificial Neural Network 46
          3.5.2   Support Vector Machine 48
   3.6   Summary...49

4. **Regional and Global Urban Land Cover Characterizations**................51
   4.1   Introduction 51
   4.2   Characterization of National Urban Land Cover..........................52
          4.2.1   Development of National Urban Land Cover Product
                   in the United States..52
          4.2.2   Updated NLCD Impervious Surface Product 54
          4.2.3   Urban Land Cover Changes in the United States
                   between 2001 and 2011 57
   4.3   Regional and Global Efforts of Mapping Urban Land 59
   4.4   Mapping Global Urban Land Cover from Nighttime Light
          Imagery...63

    4.5   Map Global Urban from MODIS Satellite Data 65
            4.5.1   Urban Extent Definition 66
            4.5.2   Classification Tool 66
            4.5.3   MODIS Data 67
            4.5.4   Training Data 67
            4.5.5   Urban Ecoregion 68
            4.5.6   Classification of Global Urban Land Cover 68
            4.5.7   Accuracy Assessment for the Global-Scale Product 70
    4.6   New Progresses on Global Urban Mapping Effort 71
            4.6.1   Remote Sensing Data 72
            4.6.2   Image Classification 72
                      4.6.2.1   Landsat Data 72
                      4.6.2.2   TerraSAR-X Data 73
            4.6.3   Multitemporal Change Detection 75
            4.6.4   Urban Footprint and Multitemporal Change Product 75
    4.7   Summary 75

5. **Assessment of Water Quality in Urban Areas** 77
    5.1   Introduction 77
    5.2   Parameters of Inland Water Quality 80
    5.3   Satellites Measurements and Empirical Algorithms for
            Different Water Quality Parameters 85
            5.3.1   Chl-a and Algal 85
            5.3.2   Colored DOM 88
            5.3.3   TSS Concentration Estimate 89
    5.4   Characterization of Pollutant Loading in a Watershed 90
    5.5   Satellite-Derived Water Quality Map 95
            5.5.1   Suspended Sediments in the Lake Chicot, Arkansas 95
            5.5.2   Water Properties in Chesapeake Bay 96
                      5.5.2.1   Satellite Chl-a Composite Images 98
                      5.5.2.2   Satellite Images and Time Series of MODIS
                                  TSS Measurements 98
            5.5.3   Water Quality Assessment for China's Inland Lake
                      Taihu 100
    5.6   Summary 102

6. **Natural Hazard Assessment for Urban Environments** 105
    6.1   Introduction 105
    6.2   Urban Development and Land Cover Transition in the Gulf
            of Mexico Region 106
    6.3   Flood Risks in Urban Environments Using Medium-
            Resolution Remote Sensing 111
    6.4   Landslide Assessment in Coastal Urban Areas 114
            6.4.1   Detection of Landslides Using Remote Sensing in
                      Hong Kong 115

6.4.2   Assessment of Landslide Environment in the Seattle
Area 118
6.4.2.1   Remote Sensing Data 120
6.4.2.2   Impervious Surface Estimate.................... 120
6.4.2.3   Evolution of Landslides in the Seattle Urban
Area 122
6.5   Summary... 127

7. Air Quality in Urban Areas—Local and Regional Aspects............. 129
7.1   Overview 129
7.2   Satellite Retrievals.. 132
7.3   Medium-Resolution Satellite Remote Sensing............. 133
7.3.1   ERS-2 133
7.3.2   Terra 133
7.3.3   Envisat 136
7.3.4   Aqua and Aura 136
7.3.5   MetOp.. 137
7.3.6   Other Platforms.. 137
7.4   Species Observations... 138
7.4.1   Trace Gases 138
7.4.2   Aerosol Remote Sensing 140
7.5   Assessments of Air Quality in Urban Areas............... 141
7.5.1   Assessment of $PM_{2.5}$ Distribution.................... 141
7.5.2   Nitrogen Dioxide.. 143
7.5.3   Ozone.. 146
7.6   Urban Land Cover and Air Quality 147
7.7   Summary... 150

8. Air Quality in Urban Areas—Global Aspects 153
8.1   Introduction 153
8.2   Satellite Systems for Global Air Quality Assessment.................. 156
8.3   Methods of Estimating the Aerosol Anthropogenic
Component.. 157
8.4   Applications of Remote Sensing for Global Air Quality
Assessment... 160
8.4.1   Characterization of Global $PM_{2.5}$ Distribution 160
8.4.2   Intercontinental Transport of Pollutant.......... 163
8.5   Summary... 167

9. Urban Heat Island and Regional Climatic Effect 169
9.1   Introduction 169
9.2   Calculation of LST from Remote Sensing..................... 171
9.2.1   AVHRR Instrument 171
9.2.2   Landsat Imagery 173
9.2.3   MODIS 174

9.3   Quantification of UHI Using Remotely Sensed LST Products... 174
       9.3.1   UHI Effects in a Local Scale 175
       9.3.2   Analysis of UHI Using Gridded Temperature and
               Land Cover Datasets 178
       9.3.3   UHI Effect across the Continental United States 182
       9.3.4   Regional UHI Effects in Europe 185
9.4   Global Aspect of UHI Effect... 187
9.5   Climate Impacts of UHI 188
9.6   Summary.. 190

References 191

Index 213

Foreword

Continued urbanization, population growth, economic activities, and global climate change have made assessing, modeling, and understanding urban areas challenging. The need for technologies, which will enable monitoring of urban assets and managing exposure to natural and man-made risks in cities, is rapidly growing (Weng and Quattrochi, 2006). Urban environmental problems have become unprecedentedly compelling in the twenty-first century. This is not only because cities and towns are the original producers of many of the environmental problems related to waste disposal and air and water pollution but also because the impact of global climate change on urban areas is becoming a matter of growing concern (Weng, 2014). Remote sensing has been an effective technology to observe, measure, monitor, and model many of the components that comprise urban environmental systems and ecosystems cycles (Weng, 2012). Urban remote sensing has emerged as a new frontier in the Earth observation technology; the volume of literature has increased rapidly since 2000 (Weng, 2014). I am pleased that Dr. George Z. Xian has the vision and energy to write a book on this important topic. Dr. Xian is an internationally renowned expert in the field of urban remote sensing and has been instrumental in the success of the National Land Cover Database development.

In the context of global urbanization and environmental changes and having recognized the benefits of urban imaging and mapping techniques, Group on Earth Observation, an international organization for exploiting the Earth observation technologies to support decision making, calls for development of a global urban observation and information system. Group on Earth Observation decided to establish, in its 2012–2015 Work Plan, a new project Global Urban Observation and Information. The main objectives of this project are to improve the coordination of urban observations, monitoring, forecasting, and assessment initiatives worldwide; to produce up-to-date information on the status and development of the urban systems at different scales; to fill existing gaps in the integration of global urban land observations with various urban ancillary datasets; and to develop innovative concepts and techniques in support of effective and sustainable urban development (Weng et al. 2014). Since the beginning of this program, Dr. Xian has been a strong supporter and serves as a colead, working with me and many colleagues to advance the project and achieve its goals.

This book marks the thirteenth volume in the *Taylor & Francis Group Series in Remote Sensing Applications* since its initiation in 2007. It is also our fifth book on urban areas. Together, these books contribute to recent developments in the theories, methods, and applications of urban remote sensing. As designated, the series aims to serve as a reference for professionals, researchers, and scientists, as well as textbooks for teachers and students. I hope that the publication of this book will promote wider and better use of remote sensing data, science, and technology and will facilitate the assessing, monitoring, and managing of cities, the *habitat* for over half of the world's population. I congratulate Dr. Xian on this important milestone in his career.

Qihao Weng
Hawthorn Woods

References

Weng, Q. (2012). Remote sensing of impervious surfaces in the urban areas: requirements, methods, and trends. *Remote Sensing of Environment*, 117(2), 34–49.

Weng, Q. (2014). What is special about global urban remote sensing? In: *Global Urban Monitoring and Assessment through Earth Observation*, Weng, Q. (Ed.). CRC Press/Taylor and Francis Group, Boca Raton, FL, Chapter 1, pp. 1–12.

Weng, Q., Esch, T., Gamba, P., Quattrochi, D.A., and Xian, G. (2014). Global urban observation and information: GEO's effort to address the impacts of human settlements. In: *Global Urban Monitoring and Assessment through Earth Observation*, Weng, Q. (Ed.). CRC Press/Taylor and Francis Group, Boca Raton, FL, Chapter 2, pp. 15–34.

Weng, Q. and Quattrochi, D.A. (2006). *Urban Remote Sensing*. CRC Press/Taylor and Francis Group, Boca Raton, FL, p. 432.

Preface

Over the past 50 years, urban development has dramatically increased around the world. Associated with urban expansions are increases in urban populations and land use. Population growth will be particularly rapid in the urban areas of less developed regions and almost all the growth of the world's total population between 2000 and 2030 is expected to occur in the urban areas of those regions. In addition, the proportion of the population living in urban areas is expected to increase substantially by 2030 in the most developed regions. In the United States, economic growth and population boom spurred increased suburbanization, which shifted residential areas to the outlying sections of a city or to separate municipalities on the fringe of urban areas. The increasing concentration of urban populations and many of the world's largest cities being outside the highest income nations represents an important change. Urban areas are becoming more and more important for current and future human settlements.

Throughout the world, land use and land cover changes associated with increasing urbanization have had significant impacts on at the local, regional, and even global scales. Landscape and environmental changes caused by urbanizations are significant in almost all urban areas. For example, urban land uses permanently change natural lands to anthropogenic impervious surfaces. These changes also influence the status of the ecosystem by altering many biophysical characteristics such as surface temperature, soil moisture content, and vegetation coverage. In addition, human settlement not only changes natural landscapes, but also transports many chemical components to urban water and air systems. Several environmental consequences are associated with runoff in urbanized watersheds, including increases in runoff volume, loss of nutrients, and significant losses of oil and grease and certain heavy metals. Rapidly built-up land may become a threat to surface water quality when total runoff increases, and hydrologic impairment leads to erosion and sedimentation. The direct impacts of urbanization include the degradation of water resources and water quality when surface runoff transports non–point source pollutants from their source areas to receiving lakes and streams. Pollutants either dissolved or suspended in water or associated with sediment, including nutrients, heavy metals, and oil and grease, can accumulate and wash away from urban areas. Therefore, impervious surfaces in urban areas have been considered as a key environmental indicator of the health of urban watersheds and as an indicator of non–point source pollution or polluted runoff. Urban areas as human settlement centers may face air quality issues when air pollution reaches certain levels. Air pollutions including particles and chemical components emitted from surfaces can be caused by individual sources on the urban scale, as well as on the local scale.

The particles introduced into the air as solids from the surface of the Earth may have negative effects on human health. Exposure to particulate matter ($PM_{2.5}$, particles with diameters less than 2.5 μm) can cause health effects, including cancers of the lung, pulmonary inflammation, and cardiopulmonary mortality. Both natural (e.g., windblown dust, sea salt from oceans, and volcanic eruptions) and anthropogenic sources (e.g., aerosols from biomass burning, combustion from automobiles, and emission from power plants) contribute to contents of $PM_{2.5}$ in the atmosphere. The transport of particulate matter of atmospheric aerosols can cause not only local or regional but also intercontinental ill-health effects. Another phenomenon associated with urban development is urban heat island effect: air temperatures in urban areas are higher than those in the surrounding rural areas. The long-term existence of the effect influences local weather patterns and may also modify local or regional climate conditions.

As more rural lands will be transformed to urban lands when more people start residing in urban areas, it is important to understand how these environmental changes may impact our living conditions as well as the ecosystem. The progress in satellite remote sensing techniques and newly launched satellites in the past 20 years holds the promise of effective monitoring of environmental conditions in many urban areas.

Driven by societal needs, spatial, spectral, and geometrical (e.g., lidar) resolutions of satellite data have been improved by many new sensor technologies and image-processing algorithms. In addition, several satellite sensors, for example, Landsat TM and ETM+, have been systematically acquiring data for many portions of the globe since the launch of Landsat in the 1980s; thus, a rich archive is available for analyzing and monitoring urban land cover changes at local and regional scales. Furthermore, urban remote sensing has improved our understanding of the biophysical properties, patterns, and processes of urban landscapes. With these observations of several decades, remote sensing data for urban areas are becoming more important for many physical models of climatic, hydrological, and ecological processes that reveal how urban areas interact with the local, regional, and even global environments.

This book intends to present a complete and exhaustive summary of the progress and advances in assessment of urban environment using remote sensing data. Current satellite observation capacities, the use of remote sensing data to characterize urban extent and urban land cover, and the applications of satellite-derived data for urban environment assessments are introduced. The book provides academic faculties, students, researchers, and government decision makers an up-to-date reference that summarizes the state of the art in both remote sensing techniques and environmental assessment methods.

George Z. Xian
Sioux Falls, South Dakota
United States

Author

Dr. George Z. Xian has been working at the Earth Resources Observation and Science Center, Sioux Falls, South Dakota, since 1997. He has focused on using remote sensing information to investigate land use and land cover changes and urban growth on both regional and national scales. He has also combined remote sensing–derived urban and land use information with other environmental information to investigate local and regional climate change and air and water quality.

1

Introduction

Urban areas have experienced dramatic changes in both population and spatial extent over the past 50 years. More than 50% of the world's population now lives in urban and suburban areas (UNFPA 2007). According to the UN Wall Chart of Urban and Rural Areas (http://www.unpopulation.org), the world's urban population was estimated at 3.29 billion in 2007 and was expected to rise to 6.4 billion by 2050. The rural population was anticipated to decline slightly from 3.37 billion in 2007 to 2.79 billion in 2050. In 2007, 49% of the world's population lived in urban areas. The world's urban population reached 49.4% in 2007, resulting in more urban residents than rural residents in the world. The proportion of the world population living in urban areas is expected to rise to 69.6% by 2050. Much of the growth is occurring in the developing countries, where urban developments are significant. The speed and the scale of this growth continues to pose formidable challenges to individual countries as well as to the world community. Figure 1.1 illustrates variations of world population between 1950 and 2050 (United Nations 2012). Associating with these population changes are large and complex economic, social, political, and demographic changes, including the multiplication in the size of the world's economy and a shift in economic activities and employment structures from agriculture to industry and services.

In addition, more people are moving into urban areas in many developing countries. For example, in 2011, cities with fewer than 500,000 inhabitants accounted for about half of the world's urban population, amounting to 1.85 billion (Figure 1.2). Cities with populations ranging between 500,000 and 1 million were home to more than 365 million people, equivalent to 10.1% of the world's urban population. Taken together, cities with fewer than 1 million inhabitants account for 61% of the urban population. One of the most significant changes following this urban population trend has been the growth in the size and importance of cities whose economies increased and changed as a result of urbanization. Another change is the number of large cities that are now centers of large extended metropolitan regions.

Throughout the world, land use and land cover changes associated with increasing urbanization have had significant impacts at local, regional, and even global scales. Associated with urban growth is the expansion of urban land use that usually translated into a significant loss of natural resources in many countries. Urban land cover is also the physical manifestation of historical, cultural, socioeconomic, political, demographic, and natural conditions. Urban growth changes, land cover conditions, and urban infrastructure alter

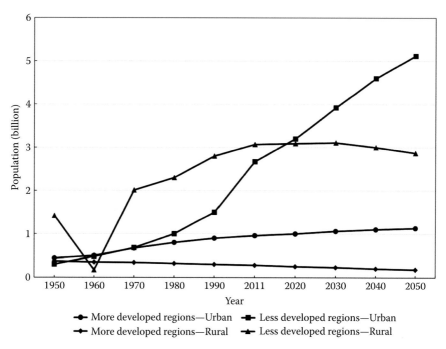

FIGURE 1.1
Urban and rural population between 1950 and 2050. (Modified from Figure 1, United Nations, *World Urbanization Prospects. The 2011 Revision*, United Nations, New York, 2012.)

surrounding environments from natural to anthropogenic types. Landscape and environmental changes caused by urbanizations are significant in almost all urban areas. The most significant change in urban landscape is from natural to anthropogenic impervious surface. Such change brings evident impacts on water resource quality, air quality, ecosystems conditions, and even micro-climate conditions. Evidence of the environmental impacts associated with urban growth have been reported by numerous researches, such as urban heat island (Oke 1973; Kukla et al. 1986; Quattrochi et al. 2000; Weng 2001; Voogt and Oke 2003), urban air quality (Nichol and Wong 2005; Lawrence et al. 2007; Xian 2007; Gurjar et al. 2008; Kanakidou et al. 2011), water quality (Ritchie et al. 2003; Carlson 2004; Jacobson 2011), and regional climate change (Landsberg 1981; Arnfield 2003; Shepherd 2005; Ren et al. 2011; Gago et al. 2013; Hebbert and Jankovic 2013; Janković 2013). Many new developments in urban areas require more pavements, which usually cause more surface runoffs. The run-off from impervious surfaces raises serious environmental concerns because of its impacts on surface water balance. Pollutants loaded with surface storm water runoff usually cause water quality problems and impact water supplies in many metropolitan areas (Matthews 2011). Generally, the impact of urban land use on environmental quality is much greater than its spatial extent would imply. Furthermore, urban land cover is an important component of

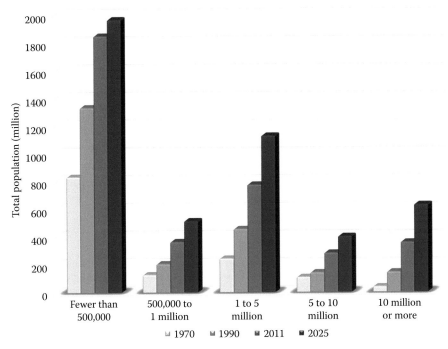

FIGURE 1.2
Total urban population in millions by city size class in 1970, 1990, 2011, and 2025. (Modified from Figure 2, United Nations, *World Urbanization Prospects. The 2011 Revision*, United Nations, New York, 2012.)

regional and global environment. Urban growth can have significant impacts on ecological, biophysical, social, and climate conditions (Imhoff et al. 2004; IPCC 2007; Seto and Shepherd 2009).

A successful urban center has had to adapt to environmental conditions and available resources, although local resource constraints have often been overcome by drawing on resources from elsewhere. The growth of urban population over the latter half of the twentieth century has also caused a very large anthropogenic transformation of terrestrial biomes, although urban centers cover only a small proportion of the world's land surface (Schneider et al. 2009). However, their physical and ecological footprints are much larger compared to their spatial extent.

Urbanization can generate both immediate and long-term influences on ecosystems in both obvious and subtle ways, often leading to unique biophysical characteristics in urban ecosystems (Alberti et al. 2003). For instance, urban development can change surface vegetation condition, and the effect of urbanization on vegetation cover depends on the form of urbanization and climate region. Low-density housing development may increase vegetation fraction and potential carbon uptake if lawns and gardens replace agricultural fields. Urban land use changes and changes in the number of people living in urban areas result in serious environmental and social problems and accelerate global environmental change (Grimmond 2007). The net ecological impact of urban centers includes the decline in the share of wild and seminatural areas. It has led not only to a decrease in biodiversity but also to fragmentation in much of the remaining natural areas and a threat to the ecological services that support both rural and urban areas (Revl et al. 2014).

Projections (Seto et al. 2012) suggest that, if current urbanization trends continue, global urban land cover will increase by 1.2 million km² by 2030, nearly tripling global urban land area between 2000 and 2030. This would mean a considerable loss of habitat in key biodiversity hotspots through destruction of the green infrastructure, which is vital in helping areas adapt to climate change impacts (Seto et al. 2012); additionally, this loss increases the exposure of population and assets to higher risk levels. Monitoring urban developments and creating sustainable urban environments remain crucial issues in the future urban development agenda.

Driven by societal desires and continuous improvement in spatial, spectral, and geometrical (e.g., radar and lidar) resolutions in satellite and sensor technology as well as image processing algorithms, remotely sensed information has been widely used in terrestrial ecosystem research and application. Urban remote sensing has evolved from very coarse-resolution interpretation to high-resolution characterization. Remote sensing is widely used among urban planners for extracting biophysical information about the urban environment, including land cover and land use mapping, urban morphology description and analysis, vegetation distribution and characterization, hydrography, and disaster relief. These data are also widely used in the field of natural resource exploration and management. The use of this type of remotely sensed data is supported by the hypothesis that the surface appearance of a settlement is the result of the human population's social and cultural behavior and interaction with the environment, which leaves its mark on the landscape (Patino and Duque 2013).

The use of remote sensing data is usually more suitable for measuring and monitoring urban environmental conditions than for urban planning purposes because in the latter case, governmental and private sector data are more easily obtained (Miller and Small 2003). However, new developments and applications have taken advantage of the consistency of the remote sensing data to extrapolate and monitor the spatio temporal dynamics of urbanization. One of the most important requirements for detecting urban and suburban features from the remotely sensed imagery is the spatial resolution of the data (Jensen and Cowen 1999). In the last decade, we have witnessed a large increase in the use of very-high-resolution space-borne sensors and programs and an increase in the availability of imagery at very high spatial resolutions. Thus, the spatial-resolution requirement of remote sensing has been met for urban land cover characterizations in many areas around the world. The open data policy for Landsat imagery, providing nearly 40 years of continuous global observation, has opened new avenues for understanding ecological and land cover dynamics (Wulder et al. 2012). The technological advancements have created a clear opportunity for both regional and global science communities to explore and use remotely sensed data for a wide range of applications.

This book focuses on introducing the latest progress in urban remote sensing and technologies used to monitor urban land use and land cover conditions and urban environment using remote sensing technology. The progress in assessing urban environment variations associated with urban land cover change, including microclimate condition and water and air quality, are also introduced. Chapter 2 introduces several methods used to extract urban landscape features using high-resolution imagery. Chapter 3 focuses on introducing urban landscape and its change assessment using medium-resolution satellite images. The methods currently used for assessing urban vegetation, impervious surface, and urban land use and land cover conditions are introduced. Chapter 4 describes how to use multi temporal satellite images to monitor urban growth around the world. Chapter 5 presents progress in applications of satellite imagery for water quality assessment in urban areas. Chapter 6 introduces use of satellite remote sensing data for natural hazard assessments in urban areas. Chapter 7 demonstrates use of medium-resolution satellite remote sensing information for air quality assessment in local and regional aspects. Chapter 8 examines global applications of satellite images for air quality analysis in major metropolitan areas around the world. Chapter 9 describes how satellite remote sensing data are used in the study of urban heat island and regional climate effects.

2

Characteristics of Urban Landscape Feature in Local and Regional Scales

2.1 Introduction

In recent years, remote sensing has become an increasingly important data source to support urban planning and management due to the availability of very-high-resolution images. Some of these images have a spectral resolution of less than 0.5 m. Such images allow extraction of detailed information on various targets in urban areas with the help of object-based image analysis. In this chapter, several methods used for characterization of high-resolution remotely sensed images for urban areas are introduced. Section 2.2 discusses the general role of remote sensing data for the urban landscape. Section 2.3 outlines the application of high-resolution satellite imagery. Section 2.4 illustrates an early effort using QuickBird imagery to map urban land cover. Section 2.5 introduces the decision tree model approach for urban land cover classification. Section 2.6 explains a recent effort using object-based image analysis for urban land cover classification. Section 2.7 summarizes this chapter.

2.2 Role of Remote Sensing for the Urban Landscape

The rapid development of worldwide urbanization and associated issues such as loss of nonurban land and wide urban sprawl addresses the need for information regarding the spatial extent of urban land, spatial distribution of various urban land uses, housing characteristics, and intensity of built-up land. The key to successfully survive the rapid urbanization in a large spatial extent is to provide accurate and timely spatial information that will assist decision makers in understanding, managing, and planning the continuously changing environment.

Traditional methods for gathering demographic data, censuses, and maps using limited samples are impractical and unsatisfactory for urban management purposes. However, remote sensing as a technique for observing the surface of the Earth from different platforms and geographic information systems can help by providing up-to-date spatial information. The synoptic view of cities afforded by satellite imagery offers great potential for data collection over urban areas. With some major improvements in remote sensing technology over the past 30 years, it is appropriate to examine how remote sensing can now be better implemented to assist urban planners and managers, and what issues the research and academic community needs to focus on to further its practical application.

Space-based remote sensing started in the late 1950s with the launch of the first military intelligence satellite. A few years later, the first U.S. meteorological satellite was launched, which was designed to aid in the production of weather observation and forecast maps. The first Earth observation satellite was launched a decade later in 1972 and is well known today as Landsat (USGS 2002). The focus of remote sensing research and application has shifted to the use of imagery acquired by Earth-orbiting satellites from using aerial photography as a tool for urban analysis because of the lower costs and frequency of updates of this imagery (Patino and Duque 2013). The earliest launched satellites are usually called the *first-generation sensors*, which were able to acquire images of the Earth's surface with relatively moderate spatial resolution, such as Landsat MSS, which had 80 m of pixel size. These images were used mainly for regional-scale studies. The second-generation satellites, such as Landsat Thematic Mapper and SPOT-high-resolution-visible (HRV), increased the spatial resolution to 30 and 10 m in visible and infrared bands, respectively, and enabled more detailed studies for the urban environment. The third-generation satellites with very high spatial resolution (5–0.5 m), such as IKONOS and QuickBird launched after 2000, have stimulated the development of newer detailed-scale applications related to the urban studies and applications. Currently, most urban environment-related studies are undertaken using both second- and third-generation satellite data to address regional and local issues in urban areas.

Generally, most of these satellites launched in the last 40 years focused on the Earth's land surface and is also called *land remote sensing*. Land remote sensing, which was developed to provide synoptic views of the surface of the Earth, has several advantages for the urban environment study. These sensors from different satellite systems can acquire images that cover a large spatial extent and provide a view that is able to identify surface objects, land cover patterns, and land cover changes. This important perspective is valuable to the interdisciplinary research that usually involved natural and social sciences in the urban environment. Remote sensing also provides continuous and consistent measurements that can overcome shortages of field survey data collection. Additionally, most sensors can cover wide spectral ranges from ultraviolet to microwave of the electromagnetic spectrum that are far beyond the range of human vision. For instance, imagery collected from thermal bands of the remote sensor is widely used to estimate surface temperature, which is a vital physical element in surface energy calculation and urban thermal morphology assessment.

Another advantage of using remotely sensed data in urban environmental research and policy is that these data can be used by obtaining internally consistent measurements of physical properties at a lower cost than that of *in situ* measurements (Miller and Small 2003). The use of remote sensing data is usually more suitable for measuring and monitoring urban environmental conditions than for urban planning purposes because in the latter case, limited urban areas are involved and governmental and private sector data are more easily obtained. However, urban extent characterized from remotely sensed data is more objective and can overcome differences in urban maps based on municipal administrative boundaries, which rarely reflect the variations in land use related to the causes and effects of urban growth (Small 2005).

The unique characteristics of remotely sensed data, such as wide area coverage and repeat cycle, provide a means for characterizing patterns of the urban landscape and other urban processes. Characterization of urban reflectance needs to meet certain conditions in terms of spatial, spectral, radiometric, and temporal resolution of the sensor.

The 30 m resolution of the Landsat sensors and the 20 m resolution of the SPOT-HRV sensors are generally not sufficient to discriminate individual features, such as buildings, streets, and trees, within the urban mosaic. The increased spatial resolution of IKONOS or QuickBird imagery provides an opportunity to image urban areas at scales sufficient to resolve many of the individual features in the urban environment. These high-resolution sensors also make it possible to image a wide variety of urban areas worldwide for a self-consistent analysis and comparison of the reflectance properties of urban land cover.

2.3 Application of High-Resolution Satellite Imagery

Optical sensors with high resolution on operational satellites provide an efficient means for quantifying past and present distributions of human settlements as well as their physical reflectance properties. High-resolution images, such as those from the satellites IKONOS, QuickBird, Orbview, and WorldView-2, are increasingly used in urban areas. These data have different spatial and spectral features that are summarized in Table 2.1. The trend toward finer spatial resolution is demanded by some specific applications that require high spatial-resolution data for urban-related issues. Generally, most high-resolution images are characterized by a high spatial resolution and a low-spectral resolution that comprises only four bands: blue, green, red, and near infrared. The limited number of bands increases the difficulty of discriminating urban land covers with similar behaviors in

TABLE 2.1
Current High-Spatial-Resolution Satellite Systems

Company	GeoEye				Digital Global				KARI		ImageSat
	Ikonos II		GeoEye-1		QuickBird-2		WorldView-2		KOMPSATS		EROS B
Sensor/Platform	Pan	Spectral Band	Pan	Spectral Band.	Pan	Spectral Band.	Pan	Spectral Band	Pan	Spectral Band	Pan
Spatial resolution (m)	1	4	0.41	1.65	0.61	2.44	0.46	1.84	1	4	0.7
Spectral resolution (nm)	525–929	445–516	450–900	450–520	450–900	450–520	450–900	400–450	500–900	450–520	500–900
		506–595		520–600		520–600		630–690		520–600	
		632–698		625–695		630–690		450–510		630–690	
		767–853		760–900		760–900		705–745		760–900	
								510–580			
								770–895			
								585–625			
								860–1040			
Swath size (km²)	11 × 11		8 × 8		16.5 × 16.5		17.6 × 17.6		15 × 15		7 × 7

the visible spectrum, such as paved roads and rooftops of different build-ings. Reflectance from these areas varies depending on the type and age of the material and variations in the solar incidence and orientation angle. The imagery itself also has several issues, including geometrical, spectral, and mapping limits (Gamba et al. 2011). Therefore, the use of high-spatial-resolution images to automatically classify intra-urban land use and land cover is still a challenge. As pointed out by several researchers, the traditional pixel-to-pixel classification approach is not efficient when using images that capture the fine details of heterogeneous urban land cover features and usu-ally contain different classes even in a small area (Blaschke and Strobl 2001; Agüera and Liu 2009; Johansen et al. 2010; Roelfsema et al. 2010; Myint et al. 2011). Furthermore, the subpixel approach may not be appropriate for urban mapping with high-resolution image data because the approach is originally designed to identify percent distribution of different land covers in coarse-resolution imagery. The subpixel processor is based on the theory that a spa-tial average of spectral signatures from two or more surface categories or endmembers can be used to represent the spectral reflectance of the majority of the pixels in remotely sensed imagery. The subpixel tool is intended to quantify materials that are smaller than image spatial resolution (Weng and Hu 2008). Moreover, Myint et al. (2011) insisted that it might not be necessary to identify percent distribution of land covers in a small pixel (e.g., QuickBird multispectral at 2.4 m spatial resolution and QuickBird panchromatic at 60 cm spatial resolution). Also, it may not be appropri-ate to model spectral responses from ground features in fewer bands (both IKONOS and QuickBird contain four bands) to effectively quantify percent distribution of many different land cover classes, since subpixel approaches use spectra of all possible land covers in all available bands.

For some specific requirements, such as storm water management and plan-ning in urban areas, accurate information about urban land cover and urban infrastructure is usually needed. For example, an important component to storm water management is detailed geographical information on surface impermeability. Impervious surface is defined as a surface that does not allow infiltration from precipitation. Surfaces in which water does not pen-etrate, such as roads and paved parking lots, increase the amount of rain-fall runoff. Increased rainfall runoff can lead to serious flooding, as well as reduce water quality in waterways and groundwater. Different approaches have been developed to classify high-resolution imagery for urban land cover characterization. For example, Thomas et al. (2003) compared three different methods, including a combination of supervised and unsupervised classification based solely on pixel-by-pixel spectral response in the high-resolution imagery with four spectral bands, the spectral classification by applying a series of spatial models to clarify confused spectral classes based on contextual clues referred to as *raster-based spatial modeling*, and with use of image segment polygons to build decision rules for classification using a classification tree statistical approach. The spatial modeling techniques achieved the highest overall accuracy (81%). In this chapter, the two methods that have been designed to exploit high-resolution data for the analysis of urban structures are introduced.

2.4 Mapping Urban Land Cover Using QuickBird Imagery

Frequently, a high-resolution image is used for a city to characterize urban land cover and land use. Efficient and effective information extraction approaches are required to meet local planning and resource management. A simple classification method was introduced to identify urban impervious surface in a mid-size urban watershed for runoff coefficient determination in Sioux Falls, South Dakota (Thanapura et al. 2007). This method chooses a simple way of classification to map urban land cover as urban and nonurban lands by analyzing the QuickBird image. The fundamental concept of this method is that normalized difference vegetation index (NDVI) generating from high-spatial-resolution imagery can be used to separate urban impervious surface from open space if appropriate thresholds are selected.

Data used include orthophoto mosaics with 0.6 m resolution, QuickBird satellite image. The orthophoto was used as a reference and the cloud-free QuickBird image was registered to the orthophoto with ground control points. Two digital orthophotographs were acquired on April 23, 2004 and May 20, 2002, respectively. Both orthophotos were collected and scanned and 90% of all contours would be within one-half of the contour interval, and 90% of horizontal positions would be within 1/30 of one inch at the specified map scale. Both orthophotos were provided in universal transverse mercator (UTM) projection, WGS84 datum, and were subset to the study area. In addition, the 2002 orthophoto was degraded 3×3 using ERDAS IMAGINE. Overlaying the subset orthophotos demonstrated good alignment between images and an image-to-image registration was not conducted.

The 11-bit QuickBird leaf-on image was collected in April 2004. The image was radiometrically corrected and orthorectified in unsigned 16-bit data type with UTM projection, WGS84 datum, and units in meters. The QuickBird image has four bands, including blue, green, red, and near infrared, and was acquired at a 2.39 m spatial resolution. The image was observed as cloud free and was registered to the 2004 orthophoto with 60 ground control points in a root mean square of 0.4846 pixel. The red channel (band 3, with a spectral range of 630–690 nm) and the near infrared channel (band 4, with a spectral range of 760–900 nm) of each registered subset image were processed to create the QuickBird (QB) NDVI imagery [NDVI = (Band 4 – Band 3)/ (Band 4 + Band 3)] for the study area. The relatively simple NDVI was utilized because it reduced heterogeneous spectral radiometric characteristics within land cover surfaces portrayed in a high-resolution image and normalized potential atmospheric effects within the image. NDVI has relatively high values for vigorously growing healthy vegetation, which has low red-light reflectance and high NIR reflectance. Impervious surfaces (e.g., asphalt, concrete, and buildings) and bare land (e.g., bare soil, rock, and dirt) have similar reflectance in the red and the near-infrared bands, so these surfaces will have NDVI values near zero.

A sequence for digital processing and analysis was proposed for mapping impervious surface areas and open spaces using 8-bit and 16-bit QuickBird NDVI imagery and developing geographic information system spatial modeling. A 50/50 spectral cluster threshold was applied within NDVI using the unsupervised classification method to maximize control over the menu of informational classes and to maximize correlation between spectral homogenous classes and the informational categories.

Decision rules have been used to characterize QB imagery to two major land cover types: impervious surface and open space. These rules have been enhanced by using NDVI data and can be implemented using almost any commonly used image processing software and can be expressed as follows:

For impervious surface

- If the land area contains less than 25% of areas that are characterized as vegetative open space, define the area as impervious surface.
- If the land area has more than 74% of areas that are characterized as impervious surface including asphalt, concerts, and building constructions, define the area as impervious surface.
- If the land area contains more than 74% of areas that are characterized as bare land, such as bare rock, gravel, silt, clay, dirt, and sand, or other earthen materials, define the area as impervious surface.

For open space

- If land area has less than 20% of areas that are characterized as impervious surfaces, define the area as open space.
- If land area contains more than 75% of areas that are characterized as natural vegetation or planted vegetation, such as grass, plants, trees, shrub, and scrub, define the area as open space.
- If land cover does not meet two criteria above, the area is defined as impervious surface.

These decision rules are simple and can be easily implemented to classify impervious surface and further define urban land cover. Furthermore, the classification scheme was used to generate QB NDVI thematic maps and to assign unsupervised spectral classes of 2, 100, and 200 spectral clusters. The same 50/50 spectral cluster threshold was then used to assign spectral clusters into two classes: impervious area and open space. For the impervious surface area, for instance, the rule requires the land area has less than 20% covered with areas characterized by a vegetative open space.

The 2004 Orthophoto and the QuickBird-derived NDVI, the QuickBird multispectral image, were used to extrapolate impervious surface and open space. The overall accuracies for both urban and nonurban lands were above 91%.

2.5 Decision Tree Model for Urban Land Cover Classification

High-resolution aerial photos are another good data sources for urban land cover classification. Digital orthophoto quarter quadrangles (DOQQs) produced by the U.S. Geological Survey (USGS) have been used to produce urban land cover datasets of 1 m resolution, which have been used to serve as training datasets to produce the USGS National Land Cover Database impervious surface product. Each DOQQ scanned from color infrared photographs acquired from the National Aerial Photography Program comprised three colors—green, red, and near infrared—with a nominal spatial resolution of 1 m. To classify urban land cover from a DOQQ image, six broad land cover classes, for example, impervious surface, trees, grass, water, barren, and shadow, were differentiated using an unsupervised classification method or supervised classification approach by manually selecting samples. The preliminary classification was further refined by screen digitizing and recoding to achieve specified classification accuracy. Then, the classification was organized to a training data file following decision tree model format. The decision tree algorithm employs attribute values of a case to map it to a leaf designating one of the classes. Every leaf of the tree is followed by a cryptic (n) or (n/m) structure. For instance, the last leaf of the decision tree is compensated (180/20), for which n is 180 and m is 20. The value of n is the number of cases in the name file that are mapped to this leaf, and m (if it appears) is the number of them that are classified incorrectly by the leaf. A nonintegral number of cases may appear because, when the value of an attribute in the tree is not known, regression tree model splits the case and sends a fraction down each branch. After decision tree models are created, they can be used to extrapolate the entire DOQQ or other high-resolution imagery into different land cover classes. Figure 2.1 illustrates urban classification using DOQQ in the Tampa Bay and Seattle areas (Xian 2007, 2008). In the Tampa Bay area, many residential houses are mixed with vegetation from DOQQ image (Figure 2.1a). The classification separates urban constructers from other land cover types (Figure 2.1b). All urban areas are displayed as pink. Vegetation and water are shown in green and blue, respectively. Other land cover types, mainly bare ground

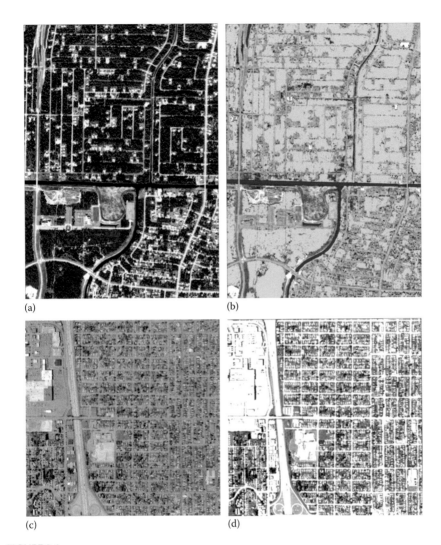

(a)   (b)

(c)   (d)

FIGURE 2.1
DOQQ in Tampa Bay (a), urban land cover classification (b), orthoimage in Las Vegas (c), and urban class for the Seattle area (d). In the Tampa Bay, land cover classification, green color represents vegetation cover, blue represents water, and urban area is displayed as pink. In the Seattle area, urban area is represented by white.

mixed with vegetation and shadows, are not colored in order to show these land covers. In the Seattle area, the 0.6 m orthoimage shows road, commercial constructers, residential housing, and trees (Figure 2.1c). Urban areas are colored as white in the classification image (Figure 2.1d). The classification in both Tampa Bay and Seattle areas clearly identify urban constructers from high-resolution images.

2.6 Object-Based Image Analysis for Urban Land Cover Classification

It is expected that more details about urban land cover features could be extrapolated from high-resolution optical data. To facilitate urban land cover and land use mapping, an object-based classifier approach has been used to identify high-resolution images, mainly of urban areas (Herold et al. 2003; Thomas et al. 2003; Hofmann et al. 2008). This approach allows analysts to use their knowledge about the environment and help to develop models that represent objects in the environment and their relationships. Recently, a similar approach was used to identify urban classes (Myint et al. 2011) and to classify land cover in an intra-urban environment (De Pinho et al. 2012). In this section, we have introduced the fundamental procedures of the method.

Generally, the object-based classification prototype starts with the generation of segmented objects at multiple levels of scales as fundamental units for image analysis without introducing per-pixel basis classification at a single scale. The object-based paradigm also differs from subpixel classifiers in that it does not consider spectra of different land covers that would quantify percent distribution of these land covers. To conduct the object-based classification, the image segmentation is performed first. Figure 2.2 illustrates a flowchart for the object-based image analysis to extract urban landscape features. Two preclassification procedures are required, including parameter and scaling factor selections, for image segmentation.

Segmentation partitions an image into different polygons or separated and homogeneous regions (objects), named *segments*, according to some homogeneous characteristic. Thus, segments contain more spectral information than single pixels, including mean, minimum and maximum band values, mean ratios, and variances. Much of the segmentation work, referred to as *object-based image analysis*, has been performed using the software eCognition (Aguilar et al. 2013). The segmentation approach consists of a bottom-up region-merging technique starting with one-pixel objects. In the following numerous iterative steps, smaller image objects are then merged into several larger ones. The multiresolution segmentation algorithm is an optimization procedure that, for a given number of image objects, minimizes the average heterogeneity and maximizes their respective homogeneity. Therefore, the outcome of this segmentation algorithm is determined by three main factors, including (1) the homogeneity criteria or scale parameter that governs the maximum allowed heterogeneity for the resulting image objects, (2) the weight of color and shape criteria in the segmentation process, and (3) the weight of the compactness and smoothness criteria. Therefore, the higher the compactness weight, the more compact the image objects may be. The optimal determination of these three terms can be carried out in a traditional way, in which a systematic trial-and-error approach validated by the visual inspection of the quality of the output image objects would be

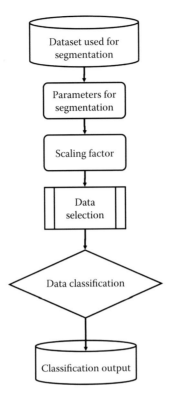

FIGURE 2.2
A flowchart of object-based image analysis for extracting urban landscape features including vegetation, building, other impervious surfaces, and water bodies.

conducted for setting these parameters (Mathieu et al. 2007). Furthermore, the configuration of the optimal parameter depends on the desired objects to be segmented. Recently, new tools have emerged for a fast estimation of the scale parameters of multiresolution segmentation (Drăgut et al. 2010), as well as for evaluating the final segmentation accuracy (Marpu et al. 2010).

In the urban classification study for Phoenix, Arizona, Myint et al. (2011) used eCognitionTM software to assign appropriate values for three key parameters, including shape, compactness, and scale to segment objects or pixels having similar spectral and spatial signatures in an image. Weights ranging from 0 to 1 for the shape and compactness factors that control the homogeneity of objects were chosen to determine objects at different level of scales. The shape factor adjusts spectral homogeneity versus shape of objects, whereas the compactness factor, balancing compactness and smoothness, determines the object shape between smooth boundaries and compact edges. The compactness or smoothness is effective only when the shape factor is not zero. The scale parameter that controls the object size that matches the user's required level of detail is the most crucial parameter of image segmentation. Numbers in the scale function determines levels of object size. Usually, the higher number of scale generates larger homogeneous objects. The selection of the level of scale depends on the size of object requires to be classified. The shape parameter was set to 0.1 to give less weight on shape and give more attention on spectrally more homogeneous pixels for image segmentation. The compactness parameter and smoothness were set to 0.5 to balance compactness and smoothness of objects equally for seven urban land covers, including building, unmanaged soil, grass, other impervious surface, pool, tree/shrub, and lake and pond. Four different scale levels to segment objects: 10, 25, 50, and 100 were employed.

The nearest-neighbor classifier uses a set of samples that represent different classes to assign class values to segmented objects. The classifier consists of two steps: (1) teach the system by giving it certain image objects as samples and (2) classify image objects due to their nearest sample neighbors in their feature spaces. The classifier is a nonparametric rule and is therefore independent of a normal distribution. The nearest-neighbor approach allows unlimited applicability of the classification system to other areas with only the additional selection or modification of new objects (training samples).

Different group of urban classes or individual classes are classified separately using different sets of parameters, different feature space (different bands, indices, or composite bands), different level of scales, and different classification rules. Here, two classification examples are introduced: buildings and other impervious surface.

The procedures displayed in Figure 2.2 can be used to extract buildings, other impervious surfaces, (e.g., road, parking lots, and specific rooftops), grass/shrub and trees, lakes and ponds, and swimming pools. To identify buildings, expert knowledge in the membership function classifier is needed. In the selection of segmentation parameter procedures, parameters were selected such as shape factor = 0.1, compactness = 0.5, and smoothness = 0.5. The scale level is level 2 = 25. Data used included mean, ratio principal component analysis (PCA), and NDVI. The analysis of membership function determines that scale-level 2 (scale parameter = 25) is the optimal scale for the classification. Also, the selected bands for the feature space include the mean of the original band 1, ratio of PCA band 3, and NDVI image. Digital values of the ratio of the PCA band 3 between 0.45 and 0.58, which intersect with a digital number (DN) value of the mean of the original band 1 higher than 390, can be used to determine buildings and vegetated areas, especially grassy features. To exclude vegetated areas, the above output is intersected with pixels having NDVI values less than 0.1. The following is the expert system rule that was employed to extract buildings in the study area.

- Ratio PCA band 3 DN magnitudes < 0.45
- Ratio PCA band 3 DN magnitudes > 0.58
- Mean band 1 DN magnitudes > 390
- NDVI magnitudes < 0.1

The output is the class of white and gray buildings. To classify other impervious surfaces including roads, driveways, sidewalks, and parking lots, the expert system rules are different from other buildings. The rules employed for the extraction of roads is the reverse of the approach employed above for identifying buildings. However, the same procedures as illustrated in Figure 2.2 were used. Hence, the same segmentation parameters as listed above were used here. However, the object scale levels were chosen as level 3 = 50 and level 4 = 100. The selected bands for the feature space still include mean of the original band 1, ratio of PCA band 3, and NDVI image. The expert system rule employed can be described as

- Ratio PCA band 3 DN magnitudes between 0.45 and 0.58
- Mean band 1 DN magnitudes < 390
- NDVI magnitudes > 0.1

The output from the classification procedure is another impervious surface.

Other urban land cover classes are characterized in the similar procedures but different rules and thresholds. For example, classes of tree, shrub, and other vegetation were extracted by the same procedures illustrated in Figure 2.2 except that a nearest-neighbor rule was chosen in the classification and different data were used as mean 2, 3, 4, ratio PCA 1, and NDVI. To determine lakes and ponds, scale level 3, and mean PCA 1, ratio PCA 3 were selected. The classifier follows the rules as

- Mean PCA and band 1 DN magnitudes < 4000 (or)
- Ratio PCA band 3 DN magnitudes > 0.62

Swimming pools have different spectral features as other water bodies. Therefore, different rules were applied to classify these objectives. An expert system rule at the scale level of 1 with a scale parameter of 10 was used to extract swimming pools in the study area. The selected bands for the feature space include a mean of PCA band 2 and mean of PCA band 3. Also, the classifier follows the rules as

- Mean PCA band 2 DN magnitudes < 15,000 (or)
- Mean PCA band 3 DN magnitudes < 24,000

Furthermore, the area contains unmanaged soil that is displayed in false color composite of near infrared, red visible, and green visible bands in red, green, and blue appeared as yellow, orange, or brown color is often confused with some rooftops that appears as same colors in the same false color display. The similar procedures illustrated in Figure 2.2 were used to extract unmanaged soil and other land cover types. Specifically, the scale level of 1 with a scale parameter of 10 was selected with the nearest-neighbor classifier.

The selected bands for the feature space include the mean of the original band 2, mean of the original band 3, and mean of the original band 4. The outputs of classification include white, gray, and yellow buildings. In addition to urban building classes, other non-urban covers including trees and shrubs, grass, asphalt roads, and soils were also characterized.

To overlay individual layers at different levels to produce a final output map of urban land cover classes, the first geographic information system overlay function starts with the last two layers by adding lakes/ponds generated by the membership function to the last output map. Hence, a priority is given to water bodies identified with the membership function when intersecting the water output with the last output generated by the nearest-neighbor classifier. All water pixels identified in the first layer that intersect with any other classes in the second layer were identified as lakes/ponds. This output map is then overlaid with the map produced for the impervious surface to add that class. After that, buildings, trees, and grass from the first two layers generated at scale levels 2 and 1 are overlaid. The swimming pool output map is overlaid last to minimize potential signature confusion with swimming pools and some building rooftops. Different types or colors of building rooftops are then merged as building class. A conceptual flowchart illustrates a step-by-step procedure to conduct the classifier procedures (Figure 2.3).

This analysis has shown that the traditional per-pixel approaches were not very effective in identifying urban land cover classes. Segmentation procedures and scale levels employed to identify objects of different classes were found to be relevant. The object-based classifier produced a significantly higher overall accuracy. However, the analysis example reveals that it is more difficult to achieve higher accuracies for larger images when dealing with a detailed urban mapping. There is no universally accepted method to determine an optimal scale level to segment objects. Moreover, a scale level may not be suitable for all classes in an image classification. The Phoenix analysis found that the best approach to select which bands to be considered for the membership function and which scale level to employ for a particular class would be to identify the class with different options and qualitatively analyze them on the display screen as a generate and test approach. The output map needs to be assessed carefully throughout the image to find any unreasonable classification. Also, some membership functions seemed to work very well for a particular class in one part of the area, but they may not perform well for the same class in other parts. It would be a good idea to treat individual classes separately to determine which method and scale level with which feature space is potentially good to extract a class. However, for some specific cases, classification of a few classes together in multispectral bands with the use of the nearest-neighbor option could be better than individual classes separately using a membership function with a set of expert system rules. It is worth to state that the threshold values given in this study may not be applicable to mapping urban environments in other areas using the same

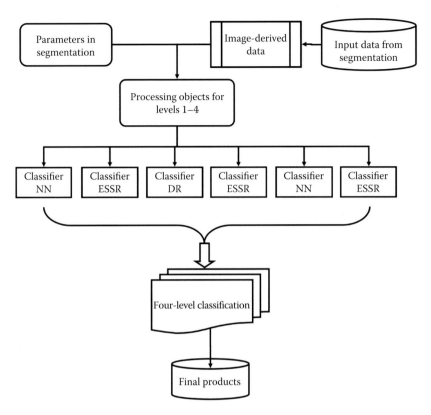

FIGURE 2.3
Conceptual flowchart showing the overall procedures for the urban land cover classification.

satellite data (QuickBird), even though the classification system employs the same classes in a similar urban environment. However, similar or considerably different threshold values with a slight modification of the parameters are expected to be effective for urban mapping in different environmental settings. Because both classifiers available in the object-based approach are based on nonparametric rules, they are independent of the assumption that data values need to be normally distributed. This method can be beneficial to a wide range of urban land cover analysis.

One of the other advantages of the object-based approach is that it allows additional selection or modification of new objects (training samples) each time, after performing a nearest-neighbor classification quickly until the satisfactory result is obtained. Many possible combinations of different functions, parameters, features, and variables are available in the object-based approach. These combinations still will not guarantee the success of mapping urban land cover using the method. The successful use of the object-based paradigm largely relies on repeatedly modifying training objects, performing the classification, observing the output, and/or testing different combinations of functions as a trial-and-error process. It needs to point out that Definiens or eCognition software is not able to perform many features or bands at many different scale levels for image segmentation and classification. Extensive use of the computer memory is common to segment tremendous numbers of objects from many different bands, especially when requiring smaller scale parameters (larger scale segmentation). This needs to be considered as a limitation, especially when dealing with a large dataset (finer-resolution data for a relatively large area). Nonetheless, the object-based classification system has been proved to be a better approach than the traditional per-pixel classifiers in urban mapping using high-resolution imagery.

2.7 Summary

Detailed, up-to-date information on urban land cover is important for urban planning and management. Differentiation between pervious and impervious land, for instance, provides data for surface runoff estimates and flood prevention, whereas identification of vegetated areas enables studies of urban microclimates. In place of maps, high-resolution images, such as those from the satellites IKONOS-2, QuickBird, Orbview, and WorldView-2, can be used after processing. Object-based image analysis is a well-established method for classifying high-resolution images of urban areas. In this chapter, several methods have been introduced for characterizing high-resolution imagery for urban areas. These studies focus on using advantages of both spatial and spectral resolutions to delineate urban landscape features. These methods have different advantages in characterizing urban cover features in different ecological environments.

3

Characterization of Urban Land Cover in a Moderate Resolution

3.1 Introduction

Remotely sensed imagery has been used as input data in regional science applications in urban settlements since the late 1950s. Thus far, moderate spatial resolution (10–100 m) imagery provides good spectral resolution with multiple bands in the visible portion of the electromagnetic spectrum, with at least one band located in the infrared portion of the spectrum and a panchromatic band. In this chapter, several algorithms that are widely used for characterization of urban landscape using medium-resolution imagery are introduced. Section 3.2 describes the most commonly used medium-resolution satellite images for characterization of urban land cover. Section 3.3 explains heterogeneous features of urban landscapes. Section 3.4 outlines algorithms to characterize urban land cover with the use of moderate-resolution imagery. Section 3.5 introduces neutral network and support vector machine (SVM) for urban land cover characterization. Section 3.6 is the summary of the chapter.

3.2 Characteristics of Urban Landscape

Satellite imagery in a moderate resolution has been widely used for characterizing urban land cover over a large scale. The spatial resolution of these moderate-resolution satellite data ranges from 15 to about 30 m in panchromatic and visible bands. Table 3.1 lists the satellite remote sensing systems that are frequently used for research and applications in urban environment. To have an overview of these satellites, Figure 3.1 illustrates the differences in spatial resolution of imagery from three different sensors of QuickBird, ASTER, and Landsat ETM+ by showing a central area in the city of Tampa, Florida. The QuickBird image reveals many details of urban structures, including roads,

TABLE 3.1

Satellite Remote Sensing Systems Used Most Often in Urban Land Cover Mapping

System	Spectral Band	Spatial Resolution	Revisit Day	Launch Time
Landsat MSS	3 visible	80	18	1972
	1 infrared			
	1 thermal infrared			
Landsat TM	3 visible	30	16	1985
	3 infrared			
	1 thermal infrared	120		
Landsat ETM+	3 visible	30	16	1999
	3 infrared	30		
	2 thermal infrared	60		
	1 panchromatic	15		
SPOT 1	2 visible	20	26	1986
SPOT 2	1 infrared			
SPOT 3	1 panchromatic	10		
SPOT 4	2 visible	20	2–3	1998
	2 infrared			
	1 panchromatic	10		
SPOT 5	2 visible	10 visible and near infrared	2–3	2002
	2 infrared			
	1 panchromatic	20 mid-infrared		
		2.5–5 panchromatic		
ASTER	3 visible	15	16	1999
	6 infrared	30		
	5 thermal infrared	90		

buildings, and trees. The ASTER image, which has a spatial resolution of 15 m in bands 1–4, also clearly shows these urban structures. The Landsat ETM+ image displays major urban structures but many mosaics, especially in many residential areas.

The use of moderate resolution remotely sensed data in the urban area is supported by the hypothesis that the surface appearance of a settlement is the result of the human population's socioeconomic activities and interaction with the environment. In urban areas, types and densities of buildings, presence and density of vegetation, types of infrastructure, and presence of remnant and natural areas can vary greatly over short distances (Cadenasso et al. 2007). The mixture of built and nonbuilt structures yields complex urban landscape configurations. In general, urban landscapes are a mixtures of several physical and biotic components including buildings, roads, grass, trees, soil, and water. However, appearances of urban landscapes that are composed of different land cover features in most moderate resolution remotely sensed images are smaller than the spatial resolution of satellite sensors. On the other hand, urban areas are generally recognized in remotely sensed imagery by

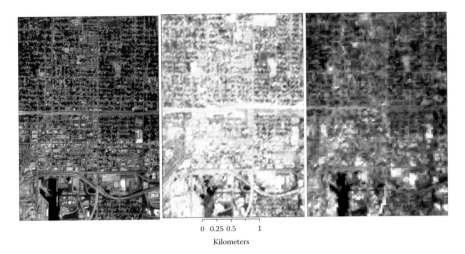

0 0.25 0.5   1
Kilometers

FIGURE 3.1
Satellite images acquired from QuickBird (left), ASTER (middle), and Landsat 7 (right) in Tampa, Florida. Images of QuickBird, ASTER, and Landsat 7 have spatial resolutions of 2.4, 15, and 30 m, respectively.

their geometric and textural characteristics. Spectral characteristics of urban land cover are less analytic than those of the nonurban land cover such as deserts and forests. There are significant differences between the spectral reflectance of urban surfaces and some nonurban lands, such as natural rock and forest surfaces, but no significant differences from other surfaces, such as bare soil and agricultural land, resulting in difficulty to detect these differences with the limited spectral resolution provided by moderate-resolution sensors such as Landsat (Small 2001, 2005). The reflectance measured by the sensor may be treated as a sum of various land cover classes of scene elements as weighted by their relative proportions (Strahler et al. 1986). The diversity of land cover types and scales in the urban mosaic therefore results in high rates of misclassification between urban and other land cover classes.

The limitations of moderate resolution remotely sensed data do not discourage research and applications of these data for the urban environment because these data have unique features of large spatial coverage, long-term and consistent data collection, and low or no cost. For example, as one of most successful Earth resource observation satellite, the Landsat program provides a continuous observation of the Earth's surface over 40 years. The Landsat multispectral scanner (MSS) sensor started to acquire data in 1972 from four spectral bands continuously from 1972 to 1992. Two of these bands (red and green) are suitable for detecting urban landscapes. The image size is about 185×185 km^2, which covers most metropolitan areas around the world within a single scene. The Landsat thematic mapper (TM) data have been available since 1982 when Landsat 4 was successfully launched. The TM sensor boarded on both Landsat 4 and 5 acquires data from seven bands, including visible, near-infrared, mid-infrared, and thermal. These spectral bands provide data that can improve the discrimination of urban land cover, particularly in areas where built-up structures are mixed with vegetation. With the improvements in both spatial and spectral resolutions, the Landsat 5 TM has continuously provided invaluable data from 1984 to 2012 for general land cover and land use mapping around the world. Landsat 7 with the ETM+ sensor onboard was launched in 1999. Before the scan line corrector failed in 2003, Landsat ETM+ provided imagery with two major improvements from TM: a panchromatic band with 15 m spatial resolution and a thermal band with 60 m spatial resolution. A new Landsat 8 with the operational land imager (OLI) and the thermal infrared sensor (TIRS) was successfully launched in 2013. Landsat 8 has been providing free of charge data since then. Tables 3.2 and 3.3 list the fundamental features of Landsat systems.

Landsat data have been widely used to characterize urban land cover and land use. In the other words, methods that use Landsat imagery to quantify urban land cover and land use represent systematic quantitative measures by using moderate-resolution satellite data in urban areas. In the subsequent sections, developments and applications that have taken advantage of the

TABLE 3.2

Landsat Systems 1–5

System	Archive Time	Spectral Resolution (µm)		Spatial Resolution (m)	Temporal Resolution (day)	Scene Size (km²)
Landsat 1	1972–1972	MSS band 4	0.5–0.6	80	18	170 × 185
		MSS band 5	0.6–0.7			
		MSS band 6	0.7–0.8			
		MSS band 7	0.8–1.1			
Landsat 2	1975–1983	MSS band 4–7		80	18	170 × 185
Landsat 3	1978–1983	MSS band 4–7, band 8 (thermal)	10.4–12.6	80	18	170 × 185
Landsat 4	1982–1993	MSS band 4–7		80	18	170 × 185
		TM band 1	0.45–0.52	30		
		TM band 2	0.52–0.60	30		
		TM band 3	0.52–0.60	30		
		TM band 4	0.76–0.90	30		
		TM band 5	1.55–1.75	30		
		TM band 6 (thermal)	10.40–12.50	120		
		TM band 7	2.08–2.35	30		
Landsat 5 (TM)	1984–1999	MSS band 4–7		80	18	170 × 185
	1984–2012	TM band 1–band 7		30–120	18	170 × 185

TABLE 3.3
Landsat Systems 7–8

System	Archive Time	Spectral Resolution (μm)		Spatial Resolution (m)	Temporal Resolution (day)	Scene Size (km²)
Landsat 7 (ETM+)	1999–present	ETM+ band 1	0.45–0.52	30	16	170 × 185
		ETM+ band 2	0.52–0.60	30		
		ETM+ band 3	0.63–0.69	30		
		ETM+ band 4	0.77–0.90	30		
		ETM+ band 5	1.55–1.75	30		
		ETM+ band 6 (thermal)	10.40–12.50	100		
		ETM+ band 7	2.08–2.35	30		
		ETM+ band 8 (panchromatic)	0.52–0.90	15		
Landsat 8	2013–present	OLI band 1	0.43–0.45	30	16	170 × 185
		OLI band 2	0.45–0.51	30		
		OLI band 3	0.53–0.59	30		
		OLI band 4	0.64–0.67	30		
		OLI band 5	0.85–0.88	30		
		OLI band 6	1.57–1.65	30		
		OLI band 7	2.11–2.29	30		
		OLI band 8 (panchromatic)	0.50–0.68	15		
		Band 9	1.36–1.38	30		
		TIRS band 10	10.6–11.19	100		
		TIRS band 11	11.5–12.51	100		

consistency of the moderate remote sensing data to study the spatial–temporal dynamics of urbanization are introduced.

3.3 Mixing Features of Urban Landscapes

Urban landscapes are composed of land cover features of buildings, roads, grass, trees, water, and soil. When moderate-resolution sensors are applied to urban areas, a direct challenge is that most urban landscape features are smaller than the spatial resolution of sensors (Strahler et al. 1986). Spatially, land cover classifications range from the general level or coarse scale that focuses on land use such as agricultural, urban, and forest to the specific level or finer scale that identifies individual components such as buildings, trees, and impervious surfaces. The coarse-scale classifications have been useful in distinguishing developed from nonurban lands across continental scales (Cadenasso et al. 2007). But these classifications cannot be used to understand the ecological function of urban areas (Pickett et al. 1997). The finer-scale classifications address the heterogeneity of urban land cover by means of classification systems that focus on individual components of the urban system. However, these classifications are limited in that they either identify only biotic components, such as parks and open lands, or identify individual landscape elements, such as walks, trees, and buildings. The finer-scale approaches reveal the richness of elements but do not deliver how those elements are arranged relative to each other.

One of the methods used in urban landscape analysis is a vegetation-impervious surface-soil (VIS) model (Ridd 1995). The model addresses the heterogeneity of urban land cover by assuming that land cover in the urban environment is a linear combination of three ecological components: vegetation, impervious surface, and soil (Figure 3.2). This model provides a general guideline for decomposing urban landscapes and creating a link for these components to remote sensing spectral data (Weng and Lu 2007). The development of digital analysis technology for remote sensing data makes it possible to implement this conceptual model for characterizing urban land cover.

The VIS model (Ridd 1995) assumes that land cover in urban areas is a line mixture of three components: water, soil, and impervious surface. The model could be applied to spatial–temporal analyses of urban morphology and biophysical and human systems. Although urban land use information may be more beneficial for socioeconomic and planning applications, biophysical information that can be directly derived from satellite data is more suitable for describing and quantifying urban structures and landscape conditions. This conceptual model has been used to quantify biophysical conditions in many urban areas, including in Salt Lake City (Ward et al. 2000), Bangkok, Thailand (Madhavan et al. 2001) and Indianapolis, Indiana

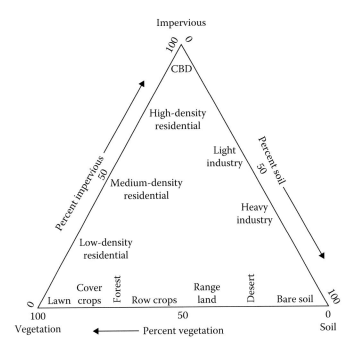

FIGURE 3.2
The vegetation-impervious surface-soil model illustrating three fundamental components of urban landscape. (Data from Ridd, M.K., *Int. J. Remote Sens.*, 16, 2165–2185, 1995.)

(Weng and Lu 2009). These studies used the VIS model to relate urban morphology to medium-resolution satellite imagery with different classification algorithms.

3.4 Characterization of Urban Landscape

3.4.1 Spectral Mixture Analysis

One of the commonly used methods for mixing landscape analysis in the urban environment is an image-based analysis procedure called *spectral mixture analysis* (SMA). SMA is used for calculating land cover fractions within a pixel and is therefore good at extracting quantitative subpixel information (Smith et al. 1990; Roberts et al. 1993; Lu and Weng 2004; Small 2005). The strength of the SMA method is based on the fact that it explicitly takes into account the physical processes detected by the observed radiances and therefore accommodates the existence of mixed pixels (Small 2003). SMA involves modeling a mixed spectrum by assuming that the spectrum measured by a sensor is a combination of spectra of all components within the pixel and the landscape can be modeled as mixtures of a few simple spectral components, called *endmembers*. SMA can be modeled as linear spectral mixture of endmember spectra, weighted by the fraction of each endmember or nonlinear spectral according to the complexity of the spectral combination.

In most urban areas, the signal received by a sensor could include reflectance from different land cover components. The moderate Landsat pixels displayed in Figure 3.1 are a mixture of vegetation, soil, impervious surface, and water. Therefore, depending on the interaction of each photon with a single land cover type within the field of view, the mixing can be treated as linear combination and the modeled spectra is the linear summation of the spectrum of each land cover type multiplied by the surface fraction they cover (Adams et al. 1986; Roberts et al. 1993; Sabol et al. 2002; Rashed et al. 2003; Wu and Murray 2003; Song 2005; Powell et al. 2007). The goal of SMA is to separate the relative proportion of endmembers. The output of SMA is a set of images that contains the fractional cover of each endmember with digital number (DN) values between 1 (100% cover) and 0 (0% cover). Figure 3.3 illustrates a general approach of SMA for characterizing urban landscape.

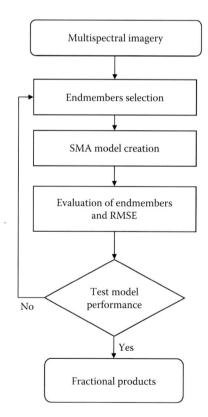

FIGURE 3.3
A general approach of SMA for characterizing urban landscape.

Generally, the linear SMA model is expressed as

$$R_i = \sum_{j=1}^{N} f_j R_{i,j} + E_i \qquad (3.1)$$

where:
  R_i is the measured value of a pixel in band i in DNs or in units of radiance
      or reflectance
  N is the number of endmembers
  f_j is the fraction of endmember
  $R_{i,j}$ is the reflectance of endmember j in band i
  E_i is the residual or the difference between observed and modeled DNs for
      band i

There are K bands in the dataset and N endmember in the mixture model. In addition, the linear SMA model is subject to the following constrain:

$$\sum_{j=1}^{N} f_j = 1 \qquad (3.2)$$

Therefore, the summation of fraction of endmembers for each pixel must be 1 (or 100% cover). Model fitness is usually evaluated by the residual term E_i or the per-pixel root mean square error (RMSE) over all image band N:

$$\text{RMSE} = \left(\frac{\sum_{j=1}^{K} E_j^2}{K} \right)^{1/2} \qquad (3.3)$$

The fraction of each endmember can be calculated by using a least squares technique to minimize the residual error E_i with the constraints on f_j. It is worth to mention that the linear mixture model may not be appropriate for applications in which only subtle spectral differences exist in all selected bands (Small 2001). In addition, the model performance also depends on the selection of endmembers.

There are several ways to select endmembers, including (1) a spectral library based on field reflectance measurements, (2) high-order endmember principal component (PC) eigenvectors, (3) manual endmember selection, and (4) the combination of image and reference endmember selection methods. For most SMA applications, the endmembers are obtained from image and can represent spectra measured at the same scale as the image data (Roberts et al. 1998). The endmembers are usually regarded as the extremes in the triangles of an image scattergram (Lu and Weng 2004). Hence, the image endmembers can be identified from the scatterplots of two spectral bands or two

PC components. For example, PC transformation can be used to guide image endmember selection because it contains almost 90% of the variances on the first two or three components and reduces the influence of band-to-band correlation (Smith et al. 1985). The eigenvalue distribution provides an estimate of the variance partition between the signal- and noise-dominated PCs of the image (Small 2005). Furthermore, the spectral mixing space represented by the multidimensional feature space of the low-order PCs can be used to describe the spectral mixtures as combinations of spectral endmembers.

However, noise variance in one band may be larger than signal variance in another band in terms of unequal scaling in different bands (Small 2001), resulting in PC transformation not being able to automatically order components according to signal information. Unlike PC transformation, the minimum or maximum noise fraction transformation orders components according to signal-to-noise ratios (Green et al. 1988; Small 2005). The minimum noise fraction (MNF) orders the PCs from high- to low- signal variance. With use of Landsat imagery, the MNF transformation usually transforms the noise covariance matrix of the dataset to an identity matrix and also conducts a PC transformation on the transformed dataset with the identity noise covariance matrix. The result of MNF is a two-part dataset, including the one associated with large eigenvalues and coherent eigenimages and a complementary one with near-unity eigenvalues and noise-dominated images. When MNF is transformed, noise can be separated from the data by saving only coherent parts in order to improve spectral processing results.

In an analysis of estimating impervious surface area conducted in Indianapolis, Indiana (Lu and Weng 2004), the MNF procedure was implemented to transform the six Landsat ETM+ reflective bands into a new dataset. The first three components accounted for the majority of the information (approximately 99%) and were used for the selection of endmembers. The scatterplots between MNF components 1, 2, and 3 are shown in Figure 3.4 and potential endmembers were selected from these components.

Four endmembers were selected corresponding to vegetation, high albedo, low albedo, and soil. Endmembers were initially identified from the ETM+ image based on high-spatial-resolution aerial photographs. The shade endmember was identified from the areas of clear and deep water, while vegetation was selected from the areas of dense grass and cover crops. Different types of impervious surfaces were selected from building roofs, airport runways, and highway intersections. Soils were selected from bare grounds in agricultural lands. After that, these initial endmembers were compared with those endmembers selected from the scatterplots of MNF1 versus MNF2, and of MNF1 versus MNF3. The endmembers with similar MNF spectra located at the extreme vertices of the scatterplots were selected. These endmembers were shade, vegetation, impervious surface, dry soil, and dark soil. A least squares regression equation defined in Equation 3.1 was then implemented to separate the MNF components into fraction images. To find the best quality of fraction images, different combinations of endmembers were tested,

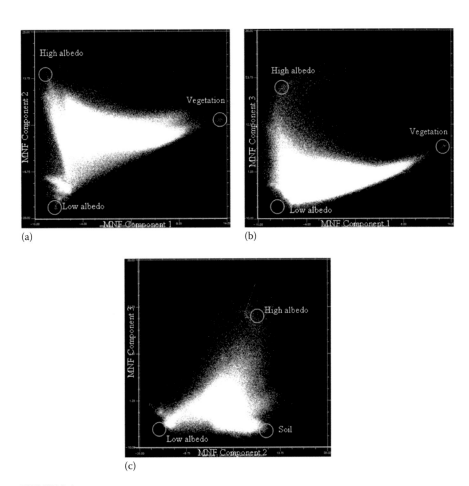

FIGURE 3.4
(a–c) Feature spaces between three MNF components. (Data from Weng, Q., and Lu, D., Subpixel analysis of urban landscapes. In *Urban Remote Sensing*, Weng, Q., and Quattochi, D.A., Eds. CRC Press, Boca Raton, FL, 405, 2007, Chapter 4, Figure 4.3.)

including four endmembers with shade, vegetation, impervious surface, and dark soil; three endmembers with shade, vegetation, and impervious surface; three endmembers with shade, vegetation, and dry soil; and three endmembers with shade, vegetation, and dark soil. Moreover, visual inspections were conducted for fraction images, analysis of fraction characteristics of representative land covers, and assessment of error images in order to determine which combination provided the best fractions.

Extra steps are usually needed to estimate impervious surface. The high albedo and low albedo endmembers are used to build a relationship to connect these endmembers with impervious surface. Through the analysis of relationships between impervious surfaces and the four endmembers, impervious surfaces are found to be on or near the line connecting the low albedo and high albedo endmembers in the feature spaces (Wu and Murray 2003). In other words, most impervious surfaces might be represented by Equation 3.1 with low and high albedo endmembers as

$$R_{imp,b} = f_l R_{l,b} + f_h R_{h,b} + E_b \qquad (3.4)$$

where:

$R_{imp,b}$ is the reflectance spectra of impervious surfaces for band b

$R_{l,b}$ and $R_{h,b}$ are the reflectance spectra of low albedo and high albedo for band b

f_l and f_h are the fractions of low albedo and high albedo, respectively

E_b is the unmodeled residual

Equation 3.4 also suggests that a pure impervious surface might be modeled by low and high albedo endmembers through a fully constrained linear mixture model. In other words, vegetation and soil endmembers have little or no contribution to impervious surface estimation. Therefore, with consideration of only three land cover types (vegetation, impervious surface, and soil) on the ground, the impervious surface fraction can be calculated by adding low and high albedo fractions. Additional procedures are usually required to take care of low reflectance materials such as water and shade and high reflectance materials such as clouds and sand.

SMA provides a practical way to quantify the VIS components of urban land cover. The method uses a regional spectral library to select endmembers to determine the fraction of each material component in the urban landscape. The subpixel fractions of specific components are grouped into general VIS classes to produce continuous maps for the three components. However, it is very difficult to automate the selection of reference endmember. Also, the method solely depends on information from reflectance of imagery and any ancillary data that could help for VIS components classification cannot be directly used in the model.

3.4.2 Regression Tree Algorithms

Another useful tool to characterize urban landscape is the regression tree model. The algorithm has been widely used to characterize land cover as continuous fields. The approach has been used to estimate impervious surface as a percent coverage in a subpixel level. One of well-known applications of the regression tree algorithm is the use of the method to produce the U.S. Geological Survey National Land Cover Database (NLCD 2011) impervious surface products. The regression method was extended by developing a classification and regression tree algorithm, which used the classification result of high-resolution imagery as the training dataset to generate a rule-based modeling for prediction of subpixel percent imperviousness for a large area (Yang et al. 2003; Xian and Crane 2005; Xian 2007, 2008). The algorithm has advantages to simplify complicated nonlinear relationship between predictive and target variables into a multivariate linear relation and to accept both continuous and discrete variables as input data for continuous variable prediction. The algorithm has been continuously used to produce the NLCD impervious surface products for the entire United States (Xian et al. 2011).

The regression tree model is a machine-learning algorithm of generating an initial tree structure model from a set of training cases (Quinlan 1993). Regression models are rule-based predictive models, in which each rule has an associated multivariate linear model. Essentially, the algorithm establishes a statistical relationship between the responding variable (dependent variable or predicted variable) and the controlling variables (independent variables) at the specific condition from training cases. A training dataset, which contains a dependent variable and numbers of independent variables, is required to constrain the parameters in each linear regression equation. A tree is grown where the terminal leaves contain linear regression models. These models are based on the predictors used in previous splits. Also, there are intermediate linear models at each step of the tree creation. A prediction is made using the linear regression model at the terminal node of the tree. The tree is reduced to a set of rules, which initially are paths from the top of the tree to the bottom. The regression tree models detect correlations of the responding variable to the controlling variables at the specific condition. The training begins with a large dataset to create rules with many nodes. The regression equation is then trained following the generated rules. The final structure of the regression tree model contains numbers of regression equations dependent on the initial number of nodes at each layer, the initial values of the weights, and the training progression. The optimally trained regression tree models can extrapolate the spatial correlations in the data while maintaining the ability to generalize fundamental features associated with the training dataset.

The condition for each rule controls the values of independent variables by different thresholds. The linear model is a simplified equation to fit the training data covered by the rule. Models based on the regression tree provide a proposition logic representation of these conditions in the form of number tree rules. Generally, the model can be expressed as *Rule* i.

If condition(s) for $x_1, x_2, x_3, \ldots, x_n$ is (are) true, then

$$y_i = a_i + \sum_{j=1}^{m} b_j x_j \tag{3.5}$$

where:

i is the ith rule

x_n are independent variables

y_i is dependent variable (e.g., percent impervious surface)

a_i and b_j are constants

m is the numbers of independent variables used in the ith rule and varies in each rule

The *i* ranged from 10 to 20 in different time. Each rule was formed according to the conditions generated from evaluating the training cases. Depending on training cases, the linear regression model can be used to calculate continuous variable such as VIS components in urban areas.

The main advantages of the regression tree algorithm include the simplifying of complicated nonlinear relationships between predictive and target variables into a multivariate linear relation and accepting both continuous and discrete variables as input data for continuous variable prediction. Moreover, different datasets that have close relationships with the dependent variable can be used as input training data.

The regression tree algorithms implemented with a data mining tool named *Cubist* (www.rulequest.com) has been used with remotely sensed data to characterize urban land cover in local and regional scales. One example of the intensive use of regression tree models is for NLCD impervious surface products that have been developed by the USGS since 2001 (Yang et al. 2003; Homer et al. 2004; Xian et al. 2006; Xian and Homer 2010). So far, impervious surface products in 2001, 2006, and 2011 have been produced for the United States using Landsat images in these years as the primary sources.

Major procedures required for implementing regression tree algorithms to estimate impervious surface in urban areas are illustrated in Figure 3.5. The first procedure is to create a training dataset. High-resolution imagery is usually used to estimate percent impervious surface cover in high spatial resolution. For example, a set of high-spatial-resolution training data was developed on imagery from the USGS digital ortho-quads and satellite-based IKONOS imagery (Yang et al. 2003). These images were classified using supervisor classification method to find impervious surface areas. Urban built-up and nonurban areas are usually labeled as 1 and 0, respectively. Then, the thematic urban and nonurban data is calculated as a percent cover of impervious surface in the pixel level. If Landsat imagery is used for large area impervious surface estimate, the percent cover is calculated in the 30×30 m^2 pixel. Figure 3.6 is an example of 0.3 m resolution orthoimage, urban and nonurban classification, percent impervious surface in 1 m pixels, and percent impervious surface in 30 m pixels in Las Vegas, Nevada.

Multiple datasets derived from high-resolution images usually are required to provide adequate training samples for a large area modeling. Generally, the number of high-resolution image selected for training dataset depends on the size of study area. Usually images that can represent major urban land cover features from different locations are selected. However, there is no specific requirement for the maximum number of images needed for the training data. If the mapping area is within a Landsat path/row footprint, several high-resolution images that represent basic features of urban land cover in the extent of the footprint are usually needed to provide sufficient training information for the regression tree models. For instance, eight 0.3 m orthoimages acquired from eight different locations in the Seattle metropolitan area were selected for mapping impervious surface in the region. Each Seattle orthoimage covers

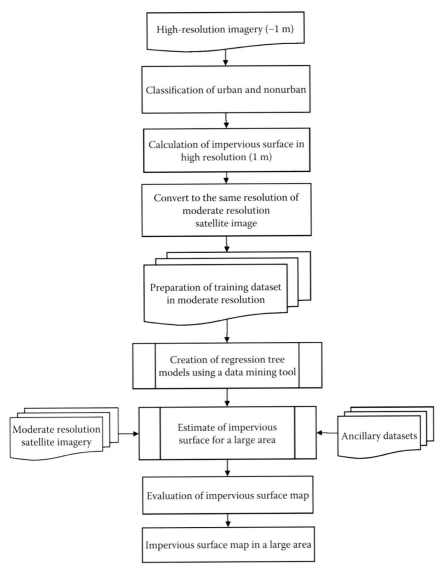

FIGURE 3.5
A general approach of estimating impervious surface in a large area using regression tree models.

approximately 1.6 × 1.6 km². Figure 3.7 displays locations of orthoimages and the Landsat ETM+ for the region. These orthoimages contains characteristics of urban land cover from low- to high-intensity urban land. The classification accuracies of high-resolution images were usually as high as 99%.

Other ancillary datasets, such as digital elevation model, slope, or other related information can also be added into the training dataset. For the use

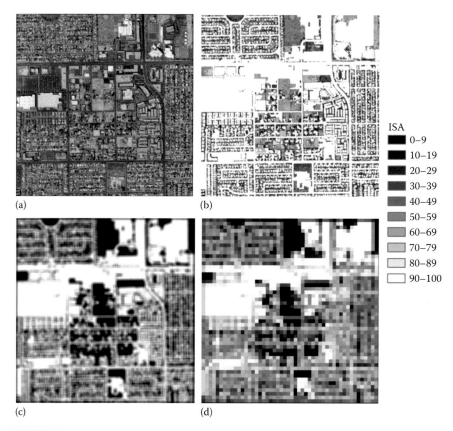

(a) (b) (c) (d)

ISA
- ■ 0–9
- ■ 10–19
- ■ 20–29
- ■ 30–39
- ■ 40–49
- ■ 50–59
- ■ 60–69
- ■ 70–79
- □ 80–89
- □ 90–100

FIGURE 3.6
High-resolution image (a), urban land (b), percent impervious surface in 1 m resolution (c), and percent impervious surface in 30 m resolution in downtown (d), Las Vegas, Nevada.

of a software Cubist (http://www.rulequest.com/cubist-info.html), which is a tool for generating rule-based models to balance the need for accurate prediction against the requirements of intelligibility, the input consists of three separate files:

- Attribute definitions (name file)
- Training dataset (data file)
- Test cases (test file)

The name file defines the attributes used to describe cases. The data file provides information from the training dataset, which Cubist will use to construct a model. The test file is optional and has the same format as the data file. The test file is used to determine the proportion of training to be used to evaluate the accuracy of a prediction model.

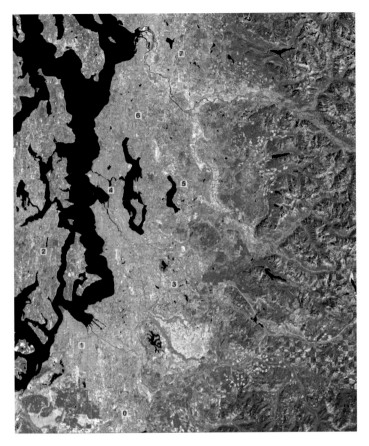

FIGURE 3.7
Locations of seven high-resolution orthoimages (labeled by number and outlined) used to estimate impervious surface in the Seattle area. The background image is a Landsat 7 image.

In the second procedure, Cubist uses the three files to construct regression tree models. As discussed previously, regression trees are constructed
using a partitioning algorithm that builds a tree by recursively splitting the
training sample into smaller subsets. In the partitioning process, each split
is made such that the model's combined residual error for the two subsets
is significantly lower than the residual error of the single best model. Each
rule includes three parts: statistical descriptions of the rule, conditions that
determine if the rule can be used, and a linear model. The statistical descriptions present the number of cases covered by the rule, the mean range of the
dependent variable, and a rough estimate of the error to be expected when
this rule is used for new data. The condition for each rule controls the values
of independent variables by different thresholds. The linear model is a simplified equation to fit the training data covered by the rule. Figure 3.8 shows
an example of a rule-based model.

Read 14797 cases (12 attributes) from 86_30k_15k.data

Model:

  Rule 1: [6287 cases, mean 0.1, range 0 to 75, est err 0.2]

```
    If
        band08 <= 294
    then
        dep = 0.1
```

  Rule 2: [2230 cases, mean 10.3, range 0 to 104, est err 13.4]

```
    If
        band01 <= 44
        band04 > 69
        band07 > 206
        band08 > 294
    then
        dep = -584.8 - 1.21 band02 + 2.1 band0.8 + 0.6
            band06 - 0.2 band05 + 0.32 band03 - 0.01 band07
```

  Rule 3: [1703 cases, mean 11.1, range 0 to 104, est err 12.5]

```
    If
        band01 <= 44
        band04 > 69
        band05 <= 85
        band06 <= 21
        band08 > 294
    then
        dep = -275.5 - 1.9 band07 - 3.05 band03 + 1.05
            band06 + 0.39 band04 + 2.4 band08 - 0.67 band01
```

  Rule 4: [2668 cases, mean 13.8, range 0 to 104, est err 13.2]

```
    If
        band01 <= 44
        band04 > 69
        band05 <= 85
        band08 > 294
        band08 <= 297
    then
        dep = -2204.1 + 8 band08 - 0.75 band07 - 0.27 band03 +
            0.06 band04
```

FIGURE 3.8

Example of regression tree models created by using Cubist. Dep is the dependent variable or percent impervious surface. Band01, band02, and so on represent independent variables, for example, Landsat bands and ancillary data, used in the training dataset.

Regression models created by Cubist are evaluated using part of data allocated in the test cases from the training dataset. Three parameters are used to evaluate model performance on training data: average error, relative error, and correlation coefficient. Similarly, the same three parameters are calculated on test data if a test file is present. Cubist also provides a scatter plot that graphs the real values against the values predicted by the model when test cases are present. Cubist also produces a prediction file that shows the actual and predicted values for every test case.

The last procedure is to make a prediction from regression equations. Models created by Cubist can be read from a free c-source code program to produce predictions from regression models. The program needs the following output files from Cubist:

- Model file
- Names file
- Data file
- Cases file

The model file contains all models produced by Cubist. The names file records all file names used for model generation. The data file is required if the model is consisted of instances-and-rules model. The cases file contains all cases that predicted values are required. At this procedure, all independent data such as Landsat imagery and ancillary are used to produce a prediction of the targeted continuous field.

For example, for a regional urban change assessment, the regression tree algorithm has been used to estimate impervious surface changes in three metropolitan areas: Seattle, Washington; Tampa Bay, Florida; and Las Vegas, Nevada, in the United States (Xian 2008). Both high-resolution orthoimages and digital orthophoto quarter quadrangles (DOQQs) were used to estimate impervious surface in 1 m resolution. A medium-resolution Landsat imagery was used to quantify impervious surface for the entire area. Eight 0.3 m orthoimages acquired from eight different locations in the Seattle metropolitan area were selected. Similarly, eight 0.3 m orthoimages from eight different locations were selected for the Las Vegas Valley. Each Las Vegas orthoimage covers approximately 1.5×1.5 km^2. Eight DOQQs covering St. Petersburg, northern St. Petersburg, Tampa, and the southern and southeastern parts of Tampa Bay were utilized to develop training datasets for the Tampa Bay watershed. For the Seattle area, Landsat scenes from path 46, rows 26 and 27, in August 14, 2002, were selected. For the Las Vegas area, one Landsat ETM+ scene was selected for path 39, row 35, from June 10, 2002. Landsat ETM+ scenes from 2002 for paths 17 and 16, rows 40 and 41, were acquired for the Tampa Bay impervious surface area (ISA) estimation. Figure 3.9 displays Landsat images used for the three study areas. Images for the three locations were acquired in clear skies for the entire areas to minimize atmospheric

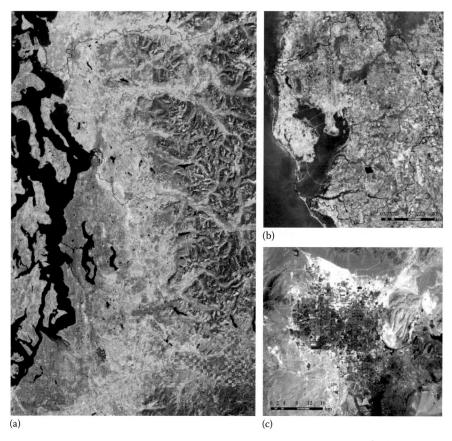

FIGURE 3.9
Landsat scenes for the Seattle area (a), Tampa Bay watershed (b), and Las Vegas Valley (c).

scatter effects. All images were preprocessed by the USGS Center for Earth Resources Observation and Science to correct radiometric and geometric distortions of the images. All images were rectified to a common universal transverse mercator coordinate system. The thermal bands had their original pixel sizes of 60 m for ETM+ images and were resampled to 30 m using the nearest-neighbor algorithm to match the pixel size of the other spectral bands. These corrections resulted in DN images that are measures of at-satellite radiance. DNs were converted to at-satellite reflectance using the following equations:

$$L_\lambda = \text{Gain}_\lambda \times \text{DN}_x + \text{Bias}_\lambda \tag{3.6}$$

where:
L_λ is at-sensor radiance
Gain_λ is the slope of radiance/DN conversion function

Bias$_\lambda$ is the intercept of the radiance/DN conversion function

Gain and bias values are provided in metadata that accompany each
Landsat image

$$\rho_\lambda = \frac{\pi L_\lambda d^2}{ESUN_\lambda \times \sin \theta} \qquad (3.7)$$

where:

ρ_λ is unitless at-satellite reflectance for ETM+ bands (1–5, 7)

θ is the solar elevation angle

$ESUN_\lambda$ is mean solar exoatmospheric irradiance

ρ_λ values from six reflectance bands were implemented in regression modeling to obtain large-area subpixel percent ISA

Impervious surface was estimated using the procedures illustrated in
Figure 3.5. High-resolution imagery was first classified to urban land cover
following the procedures illustrated above. Pixels classified as urban from
high-resolution imagery were then totaled to calculate impervious surface as a
percentage and the result was rescaled to 30 m to match Landsat pixels. Input
independent data include Landsat reflectance, normalized difference vegetation index (NDVI) derived from Landsat reflectance and thermal bands, and
slope information for building regression tree models. NDVI helped to discriminate urban residential land use from rural land in areas where housing
was mixed with trees and other vegetation canopy. The Landsat thermal bands,
however, were helpful for eliminating nonimpervious areas, especially at the
urban fringe of Las Vegas. Slope layers helped to eliminate steep areas that
were misclassified as urban in the mountain ranges surrounding Las Vegas
because most urban areas have developed in valleys or on the lower alluvial
flanks of the mountains. Slope also improved ISA estimates for the Seattle area.

Figure 3.10 displays distributions of percent impervious surface in 2002 in
the Seattle area, in Tampa Bay watershed, and in Las Vegas Valley. In the
Seattle area, many high-density urban areas (ISA > 60%) were found within
the Seattle metro region. The low-, medium-, and high-density urban areas
took about 14.8%, 9.4%, and 5.6% in the region. In the Tampa Bay area, many
low-density urban areas were mixed with trees and grass in the region
because its warm moist year-round climate is conducive to lush vegetation. Urban lands were widely spread in the watershed. The spatial extent
of total urban land use and the ratio of nonurban land-to-total watershed
area excluding water indicated that urban area reached to approximately
1800 km². The ratio of nonurban land-to-total watershed land was reduced
to approximately 70% in 2002. In the Las Vegas Valley, the wide distribution
of impervious surface is a reflection of how urban development expanded in
almost all directions. The spatial extent of urban land delineated by impervious surface reached to about 620 km² in 2002.

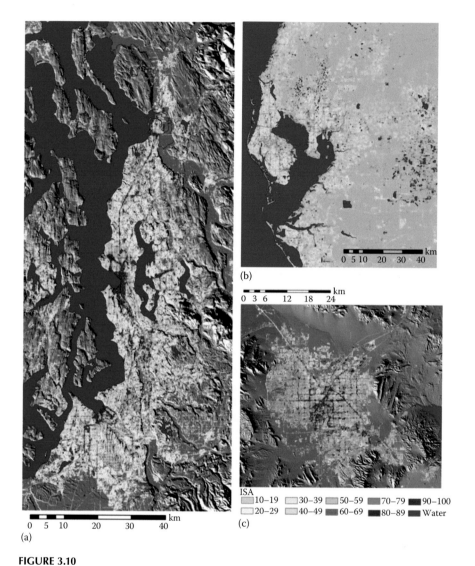

FIGURE 3.10
Percent impervious surface in 2002 in the Seattle area (a), Tampa Bay watershed (b), and Las Vegas Valley (c).

To investigate growth rates for different urban development densities, pixels were regrouped into nine categories from 10% to 100% in every interval of 10% imperviousness. The 20% to 49% imperviousness class has the largest numbers of pixels in Seattle. In addition, the largest ISA increase took place in the 20% to 59% category, or low-to-medium development densities, indicating most urban developments occurred as low-to-medium density. Categories 1–3 (10%–40% ISA) contain the largest portion of total impervious surface in Tampa Bay watershed, suggesting a large portion of urban area having buildings in relatively small size. Categories 5–7 (50%–70% ISA) contain the largest portion of total ISA in the Las Vegas area. This indicates that the largest increase of urban land use is in medium-to-high urban development densities that include most single- to two-housing units.

The regression tree model implemented for the large-area ISA estimation used several parameters to measure the model's predictive accuracy. These included average error, which is the average of absolute difference between model-predicted value and true value, and the correlation coefficient, which measures the agreement between the actual values of the target attribute and those values predicted by the model. To perform accuracy assessment using true ISA data, the modeled ISA was compared with digital close-to-true high-resolution orthoimages for selected locations. In each randomly selected orthoimage, one 5 by 5 30 m sampling unit was classified into urban and nonurban land use. A 5 by 5 30 m sampling unit registered to the Landsat pixel was defined to reduce the impact of geometric errors associated with orthoimagery and satellite images. In the 5 by 5 sample unit, areas of interest (AOI) were outlined following the boundaries of interpreted impervious surfaces. The area of each AOI was determined by using AOI property functions. The true fraction of ISA in each sample pixel was calculated by dividing the total area of AOI by each pixel area. The area classified as urban was divided by the total area of the unit, or 900 m², and taken as the true fraction of imperviousness for that unit. After all selected orthoimages were processed, comparisons were made between ISA interpreted from orthoimages and modeled ISA. Two statistics—RMSE and systematic error (SE)—were usually used to summarize differences between modeled and true impervious surfaces. The values of RMSE and SE for 7, 16, and 6 randomly selected orthoimages were calculated for the Seattle, Tampa Bay, and Las Vegas areas, respectively. The modeled ISA for Seattle area had values of 0.0 for mean SE and 15.10 for mean RMSE. ISA obtained for the Tampa Bay watershed had mean SE and RMSE values of –5.8% and 18.9%, respectively. Values of mean SE and mean RMSE were –6.0 and 16.10, respectively, for Las Vegas. The ISA estimate in Seattle had a relatively higher accuracy than that in Las Vegas and Tampa Bay, where ISA was slightly underestimated. The reason for this lower estimate for Las Vegas might be caused by the region's landscaping gravel and rocks, bare alluvial soils, and the surrounding gravel in rural areas. These materials have a similar appearance to concrete and had brighter reflectance than some urban buildings. This might cause significant confusion in regression tree models and produce lower imperviousness values for the region. In the Tampa Bay area, however, lower accuracy was associated with the regional vegetated environment. Inspection of ISA distribution indicated that the lowest accuracy came from medium-to-low residential areas where houses were usually surrounded by tall trees and other vegetation canopies.

3.5 Another Method for Urban Land Classification

3.5.1 Artificial Neural Network

An artificial neural network (ANN) is an information processing system that comprises of performance characteristics in common with biological neural networks (Fausett 1994). ANN classification algorithms have long been used for classifying remote sensing image since the late 1980s (Atkinson and Tatnall 1997). Studies have shown that ANN can produce identical or improved classification accuracies when comparing with the outcome from conventional classifiers (Civco 1993; Foody et al. 1995). The performance of ANN is contingent upon a wide range of algorithmic and nonalgorithmic parameters, such as input data dimensionality, training data, and learning processing.

There are two different types of neural network architectures: feed-forward networks and recurrent networks (Yang and Zhou 2011). The former consists of single-layer networks including an input layer that projects onto an output layer and multilayer networks having at least one hidden layer, which allows the networks to extract high-order statistics. A recurrent network differentiates itself from feed-forward networks by having at least one feedback loop that can greatly affect the training capability and performance. Various types of neural networks have been employed for image analysis, including Hopfield neural networks, multilayer perceptron (MLP), ARTMAP, and self-organizing map (SOM), but MLP and SOM are mostly used in remote sensing of impervious surfaces (Weng 2012). With consideration of neural network structures and training data performance, MLP neural networks that have no consensus on number of hidden layers, type of activation functions, or training parameters could achieve optimal performance (Paola and Schowengerdt 1995). Also, the MLP neural networks (MLPNN) are relatively easy to implement. Moreover, ANN requires fewer training samples.

The MLP network consists of three types of layers: input, hidden, and output layers. Each layer contains one or more nodes, which are interconnected to each other (Figure 3.11). They comprise distributed neurons and weighted links. In remote sensing digital-image applications, an MLP is usually arranged in an input-hidden-output layered structures. Whereas the input-layer nodes match up image bands, the output-layer nodes represent the desired land use, land cover, or surface material classes. Landsat TM image bands from bands 1 to 5 and 7 can be served as inputs to produce outputs as high albedo, low albedo, vegetation, and soil (Weng and Hu 2008). Numerous image-processing algorithms and procedures are enclosed in the hidden layer. The learning algorithm is a key to the success of an ANN model.

The design of a successful ANN model is not straightforward because the effectiveness of a model is controlled by many factors, such as the number of hidden layers, the hidden-layer nodes, the learning rate, and the momentum factor. For instance, an MLP network with the back-propagation (BP)-learning

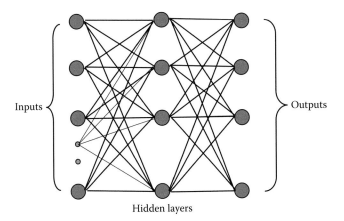

Inputs

Outputs

Hidden layers

FIGURE 3.11
Illustration of the three-layer neural-network structure used for classification. Input data can be spectral bands of imagery used for characterization. Outputs can be different landscape features or physical elements.

algorithm was implemented to estimate the percentage of impervious surfaces using Landsat imagery for Indianapolis, Indiana (Weng and Hu 2008). The ANN classifier was calculated using a weighted algorithm to calculate the input that a single node j received:

$$\text{net}_j = \sum_i w_{i,j} I_i \tag{3.8}$$

where:
  net_j refers to the input that a single node j receives
  $w_{i,j}$ represents the weights between nodes i and j
  I_i is the output from node i of a sender layer (input or hidden layer)

The output from a node j was calculated as

$$O_j = f\left(\text{net}_j\right) \tag{3.9}$$

where:
  the function f is a nonlinear sigmoidal function named as *activation function*

The input layer consists of several nodes that is determined by input data. For example, for the input Landsat image, six nodes are used to represent six reflective bands. The number of nodes of the output layer was determined by the number of land cover classes. For the impervious surface estimate, four training surface material classes were selected, including high and low albedos, vegetation, and soil. The number of hidden layer and nodes represents the complexity and capability of an ANN model. In general, one hidden layer is sufficient for most image classifications. The number of nodes in the hidden layer was estimated by the following formula:

$$N_h = \text{INT}\sqrt{N_i \times N_o} \tag{3.10}$$

where:

N_h, N_i, and N_o represent the number of nodes of the hidden, input, and output layers, respectively

In this study, only one hidden layer was used, and the number of hidden-layer nodes was four in both cases. In addition, several parameters were critical in the ANN model, including learning rate, sigmoid function constant, and momentum factor. Different values of these parameters were tested to achieve the highest accuracy.

The number of training samples is important on the effectiveness of the ANN model. Moreover, the characteristics and the distribution of samples were also important.

The samples should contain all possible spectral signatures within each class.

For instance, 30 fields were selected for each surface cover class (i.e., high and low albedos, vegetation, and soil) from the original images in the study. All samples, in which 100 pixels were randomly selected for each class to be used as training pixels, were evenly distributed in each image. Additionally, another 100 pixels were chosen as testing pixels. The accuracy rate was set to 95% because a better result can be produced with less iterations at the accuracy level. When 95% accuracy rate cannot be reached, a predefined iteration (10,000 times in this case) would stop the training process.

Impervious surface fraction image was extracted from Landsat ETM+ image. The RMSE for the result obtained from ANN was 16.7%. As a nonlinear model, the ANN model accounted better energy interactions in the urban environment where the reflected energy was very complicated. Also, the BP-learning algorithm adjusted the weight for each artificial neuron to ensure that the outputs were close to a predefined accuracy level.

3.5.2 Support Vector Machine

The concept of SVM originates from a nonparametric machine learning methodology based on structural risk minimization principle (Vapnik 2000). The basic idea of SVM is to map data into a high-dimensional space and to find the hyperplane of the different classes with the maximum margin between them. SVMs are a family of classification and regression techniques based on statistical learning theory (Cristianini and Shawe-Taylor 2000; Vapnik 2000). SVMs focus classification decisions on the boundary between classes rather than dealing with the statistical properties of classes like traditional classifiers (such as mean and variance).

SVMs map the input space (independent variables) using a kernel function to a higher dimensional space where complex nonlinear decision boundaries between classes become linear. In such high-dimensional space, an optimal linear separator is found that maximizes the margin between classes (Walton 2008). Therefore, the solution is generalized and overfitting can be reduced. The kernel function measures nonlinear dependence between the two input variables. The three important properties of SVMs can be summarized as follows: (1) the availability of a single, margin-maximizing solution; (2) product kernels for better reproducing the complex (possibly nonlinear) decision boundary; and (3) only data points closest to the separator are being played into the classification decision. Popular kernel functions include linear, polynomial, radial basis, and sigmoid kernel functions. These are the support vectors and may only be a small fraction of the training data limited to the critical area where two classes meet or overlap. Due to these inherent properties, overfitting is reduced and a manageable level of complexity is maintained as the dimensionality of the data space increases. Very large input data spaces and large training datasets (hundreds of thousands of training points) can still be handled.

SVM has been used to estimate impervious surface with the use of Landsat imagery (Walton 2008; Sun et al. 2011). Performances of different methods have been compared for impervious surface estimate. When comparing methods using all input bands, regression tree models consistently produced better results and SVR the worst by comparing MAE and RMSE (Walton 2008). However, for a study of estimating impervious surface in Beijing, China, SVM provides significantly better classification, performance than the MLPNN (Sun et al. 2011). The overall accuracy and overall kappa in SVM were 96.55% and 0.954 compared to 93.80% and 0.917 in MLPNN, respectively. For the high albedo class, the producer's accuracy and user's accuracy were both larger in SVM than those in MLPNN. The coefficient of determination shows that SVM also generated better results than MLPNN.

3.6 Summary

In this chapter, features of urban landscape and different methods for characterizing the urban land cover in moderate resolution have been emphasized. We explained the complexity of urban landscape configurations. Generally, urban landscapes are a mixture of buildings, roads, grass, trees, soil, and water. However, appearances of urban landscapes that are composed of different land cover features in most moderate resolution remotely sensed images are smaller than the spatial resolution of satellite sensors. Urban areas are generally recognized by their geometric and textural characteristics in moderate resolution remotely sensed imagery. There are significant differences between the spectral reflectance of urban surfaces and some nonurban lands, such as natural rock and forest surfaces, but no significant differences from other surfaces, such as bare soil and agricultural land, resulting in difficulty to detect these differences with the limited spectral resolution provided by moderate resolution sensors such as Landsat. For this reason, different algorithms, such as spectral mixing, regression tree, neural network, and SVM have been used to characterize urban land cover. The technical procedures introduced here were designed and implemented to extrapolate the principal source of data for urbanization studies. The use of this type of information derived by remotely sensed data can improve our understanding of the human population's social and cultural behavior and interaction with the environment, which leave their mark on the landscape. The remotely sensed data from satellites can also be used to measure the context of social phenomena, to gather additional contextual data on the environment in which people live, and to measure the environmental consequences of social processes.

4

Regional and Global Urban Land Cover Characterizations

4.1 Introduction

Driven by a constantly accelerating increase of urban population in recent decades, anthropogenic developments changed the Earth's surface in both regional and global scales. Urban sprawl has become one of the most dynamic processes in the context of global land use transformations. An increasing number of international environmental agreements place global change at the top of international scientific and political agendas, including the Kyoto Protocol, the Convention on Biological Diversity, and the Convention to Combat Desertification and the Ramsar Convention on Wetlands (McCallum et al. 2006). There are several hundreds of multilateral environmental agreements and bilateral agreements dealing with different aspects of the environment and global change. These agreements require a unique set of information for implementation, monitoring, and compliance. A key component of the data needed within the global change framework is ecosystem-based information.

The expansion of urban agglomerations is closely associated with a substantial increase of impervious surface in local, regional, and global scales (Esch et al. 2009; Xian et al. 2011; Taubenböck et al. 2012; Weng 2012). Elements of the built-up land result in at least some physical evidence in virtually all terrestrial Earth observation data. However, the spatial extent and density of a metropolitan area vary in different regions or counties. Most current multi temporal maps of urban extent are generally based on municipal administrative boundaries that rarely reflect the variations in land use related to the causes and effects of urban growth (Small 2005). The definition of urban land may vary in different countries. A systematic quantitative measure of the spatial extent and development density of urban areas would facilitate the application and research by providing comparisons of urban morphology and development intensity in different physical, cultural, and socioeconomic settings. Global to regional scale analyses of the connections between urbanized areas and natural and anthropogenic processes require high-quality, regularly updated information on the patterns and processes within the urban environment—including maps that monitor location and extent (Potere and Schneider et al. 2007). In this chapter, Section 4.2 illustrates the development of U.S. Geological Survey National Land Cover Database (NLCD) urban land cover products. Section 4.3 reviews regional and global efforts of mapping urban extent. Section 4.4 introduces quantification of global distribution using nighttime light satellite data. Section 4.5 illustrates how to use moderate-resolution imaging spectroradiometer (MODIS) data to map global urban extent. Section 4.6 introduces some recent progresses on global urban land mapping. Section 4.7 is the summary of the chapter.

4.2 Characterization of National Urban Land Cover

Urban land cover as an important component of regional and global environmental change has significant implications for a range of ecological, biophysical, social, and climatic consequences. Urban growth in the last several decades has converted a substantial amount of nonurban landscape into anthropogenic impervious surface in the United States. The systematic and consistent urban land cover information and monitoring for the nation is important for assessing the status and health of terrestrial ecosystem, understanding spatial patterns of biodiversity and developing land management policy for natural resource management, and socioeconomic assessments. Information about land cover and land use (LCLU) change is often used to produce landscape-based metrics and evaluate landscape conditions to monitor LCLU status and trends over a specific time interval (Loveland et al. 2002). However, at regional or national scales, such efforts face a number of challenges, including timely acquisition of data, the high cost of creating national products, and the development of appropriate analytical techniques to successfully evaluate current condition and associated change.

4.2.1 Development of National Urban Land Cover Product in the United States

In the early 1990s, several federal agencies of the United States recognized the need for national land cover of medium spatial resolution to support their environmental and natural resource management programs. Multiresolution land characteristics consortium (MRLC) was convened to pool resources for the purchase and processing of Landsat data for the conterminous United States to U.S. federal agencies in a standard processed format. These data included seasonal Landsat 5 imagery mosaics for the nation and were then used to develop a seamless national land cover product for the conterminous United States. In 1999, the MRLC consortium initiated an expansion that resulted in six new agencies joining the consortium: National Aeronautics and Space Administration, National Park Service, Natural Resources Conservation Service, Bureau of Land Management, U.S. Fish and Wildlife Service, and the Office of Surface Mining. This expansion enabled the creation of a more comprehensive land cover database for the nation, encompassing all 50 states and Puerto Rico (Homer et al. 2004). Through an umbrella of MRLC, the U.S. Geological Survey (USGS) produced the NLCD for the entire United States.

The first generation of the national land cover was created as a 30 m resolution land cover data layer over the conterminous United States from circa 1992 Landsat Thematic Mapper (TM) imagery as a National Land Cover Dataset 1992 (NLCD 1992; Vogelmann et al. 2001). The land cover classification scheme designed for NLCD 1992 was a modified land use and land cover classification system with 21 classes using either an unsupervised or a supervised classification method. For the unsupervised method, a *k*-mean clustering algorithm was applied to either a leaf-on or leaf-off Landsat image to generate spectral clusters, with clusters labeled subsequently by interpreters. In many cases, ancillary data (e.g., census, slope, aspect, and elevation) were used to resolve spectral confusion, so that each Landsat TM pixel could be labeled into one of 21 land cover classes. A supervised approach using a classification tree algorithm was used at the later stage of the project. Three urban land cover classes including low-intensity residential, high-intensity residential, and commercial/industrial/transportation were defined in NLCD 1992 product.

NLCD 2001 was designed to incorporate many of the lessons learned from NLCD 1992 production, as well as target new MRLC 2001 member requirements. NLCD 2001 followed a database approach that moved beyond traditional remote sensing classification of land cover focusing on a single-legend classification system and a single land cover layer that meet only specific requirements. Ecologically based mapping zone delineation was used to define NLCD 2001 mapping regions that are relatively homogeneous regions with distinct ecological features. Six factors were considered in defining these mapping zones, including physiography, land cover characteristics, spectral feature uniformity of Landsat imagery, edge-matching feasibility among mapping zones, the size of each mapping zone, and the number of Landsat images required to make a mosaic in a zone. A total of 66 mapping zones were identified for the conterminous United States. Three products, including land cover, percent impervious surface, and percent tree canopy, characterized for each 30 m cell were produced for the entire nation. In NLCD 2001, urban land cover was defined as four categories using percent impervious surface product. With different thresholds of percent impervious surface, urban land cover was characterized as different development levels, including open space with $0 < $ impervious surface area (ISA) $ < 20\%$, low intensity with $20\% \leq ISA < 50\%$, medium intensity with $50\% \leq ISA < 80\%$, and high intensity with $ISA \geq 80\%$. After that, the same urban land cover definition was implemented to all NLCD land cover products.

The percentages of urban impervious surfaces were estimated at Landsat subpixel level using a regression tree algorithm, which is illustrated in Chapter 3. The algorithm generalizes a set of rules based on training data to predict percent imperviousness density for each pixel of an image. Essentially, the algorithm establishes a statistical relationship between known imperviousness at a location and the corresponding Landsat spectral reflectance. The regression tree model then uses a sequence of multivariate linear equations to calculate percent impervious surface in urban areas. Impervious surface was estimated using Landsat images as the primary data source for the entire nation.

Extensive and high-quality training data were collected for the use of Classification and regression trees (CART) in NLCD 2001. The success of urban land cover characterization using CART critically relies on the availability of abundant high-quality training data. A major effort was devoted to the creation of training data for imperviousness estimates. For modeling imperviousness, a set of high-spatial-resolution training data was developed on imagery from the USGS digital ortho-quads and satellite-based IKONOS imagery. Typically, six to eight small representative urban areas were classified as impervious and nonimpervious for each 1 m pixel on these high-resolution sources. They were then resampled to 30 m proportions for subsequent training of the regression tree models. The initial products were tested by using one or two digital orthophoto quarter quadrangles (DOQQs) of 6 to 8 km2 per Landsat path/row to classify impervious surface. Later, 3 to 5 smaller chips of 1 to 4 km2 in size were quantified throughout the Landsat scene and accuracy and prediction quality was improved. Increasing the sampling frequency improved both the reliability of training distributions to capture the total range of zonal estimates and classification efficiency with less costs. The completed training images were then extrapolated across mapping zones using regression tree models to derive continuous imperviousness estimates.

The NLCD 2001 impervious product suggests that groupings from 1%–10% represent the largest proportion at about 47.1% of total area (likely due to the large number of tertiary roads outside of urban areas), and groupings from 91%–100% represent the smallest proportion at 1.5% of total area. The total spatial extent of impervious surface is approximately 457,100 km^2.

4.2.2 Updated NLCD Impervious Surface Product

The NLCD 1992 and 2001 products have been released based on a 10-year cycle. NLCD has been widely recognized as an important data source to quantitatively describe terrestrial ecosystem conditions and support landscape change research. With these national data layers, there has been about a five-year time lag between the image capture date and product release. However, in some areas, terrestrial ecosystems frequently experience significant natural and anthropogenic disturbances during production time, resulting in products that may be out of date when it is completed. To keep the NLCD as temporally relevant as possible, a new frequency of five years was implemented to provide more timely capture of land cover change. The updated product is designed to provide the latest land cover and it change that can be used to identify the pattern, nature, and magnitude of changes occurring between two time periods.

To achieve this goal in a cost-effective manner, the selected approach seeks to identify areas of landscape change occurring after 2001 and to update land cover data only for those changed areas. For the areas that have not changed, the land cover would remain the same. The temporal sequences of remotely sensed images can accurately indicate spectral changes based on surface physical condition variations, assuming that digital values are radiometrically consistent for all scenes (Cakir et al. 2006). Therefore, multi-temporal remotely sensed data can be used as primary sources to identify spectral changes and extrapolate land cover types for updating land cover classification and the continuous variables related to LCLU change over a large geographic region.

Change vector analysis approach was used as an early effort to update NLCD 2001 impervious surface product through identifying changed pixels. Therefore, regression tree models trained by using NLCD 2001 data designed in unchanged pixels could be built up and used to calculate impervious surface as a continuous variable (Xian et al. 2009; Xian and Homer 2010). This approach, however, heavily relied on image change detection to locate changes between two times and label these changes to different land cover classes. The land cover data were then used with the ISA product to mask out nonurban land. The accuracy of final impervious surface depended on both land cover classification and imperviousness estimation, resulting in the prototype method not robust enough for the nationwide impervious surface mapping effort. To obtain an optimal estimate of ISA, it is preferable to have a method that directly relates to ISA estimation in a fast and economic manner. Therefore, an improved approach was developed with the requirement that the percent imperviousness estimated by regression models was consistent and accurate if reliable training data were used. The improved method also introduces the use of nighttime lights imagery from the National Oceanographic and Atmospheric Administration (NOAA) Defense Meteorological Satellite Program (DMSP) in addition to Landsat imagery in the modeling process to update NLCD impervious surface product for the nation (Xian et al. 2011, 2012).

Generally, the approach comprised three major procedures: creating a training dataset in step 1, modeling synthetic impervious surfaces in step 2, and comparing model outputs for optimal selection and final product clean up in step 3. Figure 4.1 illustrates these procedures. These procedures were implemented in every Landsat paths and rows that cover entire CONUS. To estimate the percent impervious surface for a target year, following data inputs were used: Landsat pairs in the same season in both baseline and target years, nighttime stable-light satellite imagery in both target and

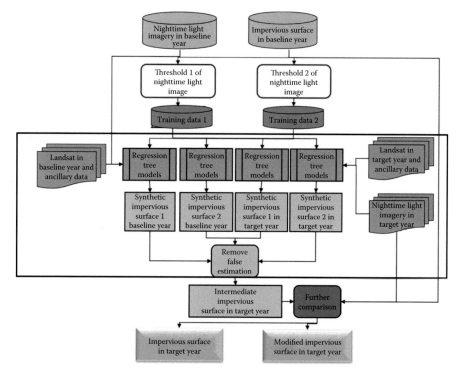

FIGURE 4.1
Flowchart of procedures to update impervious surface between a baseline year and a target year.

baseline years, and the NLCD impervious surface and land cover products in the baseline year. All Landsat images were converted to top-of-atmosphere reflectance. Nighttime light images in were converted to 30 m and subset into each Landsat path and row. For example, to update NLCD 2006 impervious surface, baseline time is 2001 and target time is 2006. Subsequently, in the procedure of training dataset creation, 2001 nighttime light imagery was first imposed on the NLCD 2001 impervious surface product to exclude low-density imperviousness outside urban and suburban centers, so that only imperviousness in urban core areas could be used in the training dataset. Two training datasets, one having a relatively larger urban extent and another having a smaller extent, were produced through imposing two different thresholds on nighttime light imagery. In the synthetic modeling step, each of the two training datasets combined with 2001 Landsat imagery was separately applied using the regression tree algorithm. Two sets of regression tree models were created to estimate percent imperviousness. Therefore, two 2001 synthetic impervious surfaces were then estimated. Similarly, the same two training datasets were used with 2006 Landsat imagery to create another two sets of regression models to produce two 2006 synthetic impervious surfaces. In the model output comparison procedure, both 2001 and 2006 synthetic impervious surface pairs were compared with use of both 2001 impervious surface products to retain 2001 ISA in the unchanged areas. The 2006 nighttime light image was then employed to ensure that non-urban areas were not included and that new impervious surfaces emerged only in the nighttime light extent. After this step, two 2006 intermediate impervious surfaces were produced. In the final composition step, the two intermediate products and 2001 imperviousness were compared to remove false estimates in nonurban areas due to strong reflectance from nonurban area with the following constraints:

$$ISA_{i,n}(2006) = \begin{cases} 0, & S_n(2001) > S_n(2006) \quad \text{and} \quad S_n(2001) \not\subset ISA(2001) \\ S_n(2006), & S_n(2001) = 0 \quad \text{and} \quad S_n(2006) > 0 \end{cases} \quad (4.1)$$

where:
  ISA(2001) is the 2001 impervious surface area from the NLCD 2001 product
  S_n(2001) and S_n(2006) are synthetic imperviousness estimates in 2001 and
      2006, respectively
  n represents the synthetic products 1 and 2
  $ISA_{i,n}$(2006) is the 2006 intermediate estimate

Finally, impervious surface in each Landsat path and row were mosaicked together to produce a seamless 2006 impervious surface product for the entire conterminous United States. The updated imperviousness product contains both new ISA emerged from the 2001 nonurban area and ISA intensified from 2001 low-density urban area using the threshold designed in our original method (Xian and Homer 2010). Recently, the algorithm was implemented to produce NLCD 2011 impervious surface products.

4.2.3 Urban Land Cover Changes in the United States between 2001 and 2011

Increases in impervious surface area associated with urban land cover expansion were directly estimated from the updated ISA product. To directly compare 2001 and 2011 imperviousness variation, the percent imperviousness products were regrouped into 10% categories. The new ISA in 2006 had a total area of 6,360 km², or a 6.39% increase from the 2001 base amount across the CONUS. The new ISA had about 4,331 km² in 2011, or 4.09% increase from the 2006 base. The growth of ISA reached approximately 972 km² per year between 2001 and 2011. The total ISA in 2011 reached 110,282 km², which is slightly larger than the state of Tennessee. The 2011 urban land cover defined as four categories described above was 454,292 km², which is close to the size of California and West Virginia. Figure 4.2 illustrates the spatial distribution of 2011 urban land cover across the conterminous United States. The 2001 and 2011 new urban land cover in three metropolitan areas of Tampa

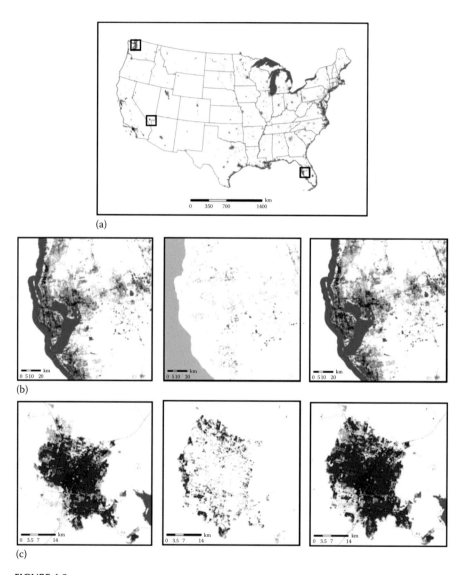

(a)

(b)

(c)

FIGURE 4.2
2011 urban land cover across the conterminous United States (a) and in three metropolitan regions: Tampa Bay (b) and Las Vegas (c).   (*Continued*)

Bay, Florida; Las Vegas, Nevada; and Seattle, Washington, are enlarged and included in the figure. Most new urban land covers emerge around existing urban areas and exhibit ring or *donut* patterns. The large *donut* expansion in both Tampa Bay and Las Vegas areas indicates the imperviousness growths are in fast and widespread patterns associated with urban development in all

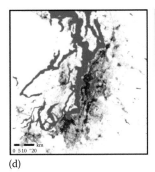 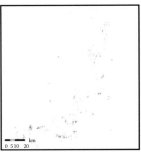 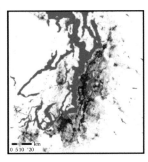

(d)

FIGURE 4.2 (Continued)
2011 urban land cover across the conterminous United States: Seattle metropolitan region (d). From left to right are urban land cover in 2001, changes of urban land cover between 2001 and 2011, and urban land cover in 2011.

directions in these regions. Alternatively, some impervious surface growth is restricted by regional topographic features. The new urban land cover growth in the Seattle area widens the existing urban strip along the valley and is restricted by mountains on the east and seas on the west.

Furthermore, characteristics of imperviousness and its change across eight economic regions were summarized. These eight regions are defined by the U.S. Department of Commerce and are illustrated in Figure 4.3. Generally, Plains and Rocky Mountains have similar distribution patterns, peaking in 20% imperviousness. The dominant fractions of ISA emerge in 60% category in the New England region, 30% category in the South East region, 50% category in the Great Lakes region, 60% in the South West region, and 70% in the Far West region. The imperviousness does not have apparent dominant pattern in the Mideast region, where impervious surfaces above 40% have almost similar spatial extents. These spatial distribution patterns are related to regional housing development patterns and urban land use densities.

4.3 Regional and Global Efforts of Mapping Urban Land

The use of satellite remote sensing is an obvious alternative methodology for measuring and monitoring the location and extent of urbanization in a large area. However, urbanization has been difficult to measure and assess globally by using remote sensing. The spectrally diverse land cover types in urbanized areas are normally mixed and confused with nonurban lands and matters handling both the spatial resolution and spectral resolution of the sensors are important.

There is a trade-off between spatial and temporal resolutions when remote sensing data is used to characterize urban landscape. The higher

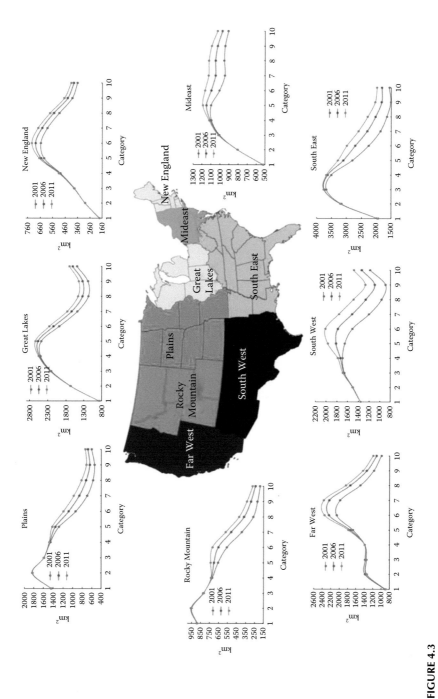

FIGURE 4.3
Regional patterns of impervious surfaces and their changes between 2001 and 2011.

spatial-resolution imagery essential for detailed analysis of urban features, including city planning, property assessments, and community development, is gathered infrequently, often contains less spectral information and is expensive. When regional or global data coverage is needed, the practical options are medium- to coarse-resolution data. All recent global land cover mapping efforts rely on coarse-resolution imagery, facilitated, in part, by the launch of second generation coarse-resolution Earth observation satellites, such as the MODIS and other sensors with improved radiometric quality, higher-temporal-resolution and higher geo-reference accuracy than earlier sensors, such as the advanced very-high-resolution radiometer (AVHRR). Although the spatial resolution of these data is relatively coarse (250–1000 m), the spectral precision and near-daily acquisition rates ensure that these datasets are capable to monitor urban areas on a regional or a global scale.

Attempts to create a global urban land cover dataset from satellite observations started in the early 1990s. Table 4.1 describes 10 global-scale maps that can be used to identify urban areas (Potere et al. 2009). The first attempt to build a global land cover map using satellite imagery was conducted by the Global Land Cover Characteristics database, a joint effort between the USGS and the European Joint Research Centre. The Global Land Cover Characteristics was accomplished by using one-year (1992–1993) observations from AVHRR (Loveland et al. 2000). The Global Land Cover Characteristics did not map urban areas directly. Instead, the digital chart of the world, a map developed from VMAP0 (Danko 1992), was used. The same approach was used by the University of Maryland group to produce the global AVHRR-based land cover map (Hansen et al. 2000). It is worth to note that VMAP0 was created by digitizing maps and navigational charts between 1950 and 1979. Because VMAP0 was the first comprehensive global dataset, it was used as part of the input stream for other global land cover maps regardless of its poor geolocations for some urban polygons.

Several studies have been conducted to evaluate the use of medium- to high-resolution satellite data for urban areas in different cities around the world. These studies range from individual cases to cross-urban area comparisons (Small 2003; Seto and Fragkias 2005; Schneider and Woodcok 2008). Meanwhile, global effects aiming at operational monitoring of the effects of spatial urbanization have produced several land cover maps by using MODIS and other moderate-resolution data. Among the 10 global-scale land cover maps, five of them—MOD500, MOD1K, GLC00, GLOBC, and IMPSA—were derived from a full year of coarse-resolution satellite imagery. MOD500 and MOD1K were produced from multiple spectral imagery from the MODIS instrument in circa 2001 (Schneider et al. 2003, 2009, 2010). MOD500 and MOD1K were based on 463 m resolution Collection 5 data and on the 927 m resolution Collection 4 data, respectively. Manual interpretation of Landsat imagery and other ancillary data approach was employed to construct training sites for supervised classification in both MOD1K and MOD500. GLC00 was also produced using

TABLE 4.1

The 10 Global Urban Maps and Their Features, Including Map Type, Resolution, and Global Urban Extent

Code	Map Name	Producer	Map Type	Resolution
VMAP0	Vector Map Level Zero	U.S. National Geospatial-Intelligence Agency (http://geoengine.nga.mil)	Thematic	1:1 mil
GLC00	Global Land Cover 2000 v1.1	European Commission Joint Research Center (http://www-gvm.jrc.it/glc2000)	Thematic	988 m
GLOBC	GlobalCover v2	European Commission Joint Research Center (http://ionial.esrin.esa.int)	Thematic	309 m
HYDE3	History Database of the Global Environment v3	Netherlands Environmental Assessment Agency	Continuous (percent urban)	9000 m
IMPSA	Global Impervious Surface Area	U.S. National Geophysical Data Center (U.S. NOAA) (http://www.ngdc.noaa.gov/eog/dmsp.html)	Continuous	927 m
MOD500	MODIS Urban Land Cover 500 m	University of Wisconsin, Boston University, U.S.-NASA (http://www.sage.wisc.edu)	Binary	463 m
MOD1K	MODIS Urban Land Cover 1 km	LP DAAC, U.S. NASA and USGS (https://lpdaac.usgs.gov/products/modis_products_table/)	Binary	927 m
GRUMP	Global Rural-Urban Mapping Project, alpha	Earth Institute at Columbia University (http://sedac.ciesin.columbia.edu)	Binary	927 m
LITES	Nighttime Lights v2	U.S. National Geophysical Data Center (U.S. NOAA) (http://www.ngdc.noaa.gov/dmsp)	Continuous	927 m
LSCAN	LandScan 2005	U.S. Oak Ridge National Laboratory (U.S. DOE) (http://www.ornl.gov/sci/landscan)	Continuous	927

Source: Potere et al., *Int. J. Remote Sen.*, 30, 6531–6558, 2009.

Note: 10 global urban or urban-related maps listed in order of increasing global urban extent. DOE, Department of Energy; MODIS, Moderate Resolution Imaging Spectroradiometer; NASA, National Aeronautics and Space Administration; NOAA, National Oceanographic and Atmospheric Administration.

one-year data in circa 2001 from the SPOT4-VEGETATION instrument. Moreover, the GLC00 map was accomplished by 18 separate production teams, each relying primarily on unsupervised classification methods (Bartholome and Belward 2005).

The GLOBC map was developed from medium resolution imaging spectrometer (MERIS) data spanning May 2005–April 2006 with 309 m spatial resolution. The unsupervised approaches similar to those of GLC00 were chosen for the GLOBC. The exception was Australia, where it appears that the road network and other map layers were incorporated into the classification process (Potere et al. 2009). GLOBC has relied on the GLC00 map directly for the urban class in South America, Western Asia, Africa, India, and Japan. The IMPSA map was produced by modeling impervious surface area with use of LSCAN2004 and LITES 2000–2001 data (Elvidge et al. 2007a). The IMPSA model was trained using exemplars from Landsat-based maps of impervious surface area for the United States (Homer et al. 2007).

The two other urban maps, HYDE3 and the global rural-urban mapping project (GRUMP), were produced by employing both remote sensing and ground-based inputs. GRUMP integrated VMAP0, 1994–1995 LITES, census data and ancillary GIS datasets (CIESIN 2004). GRUMP also used NOAA's nighttime lights dataset to delineate urban areas. HYDE3 combined LSCAN 2005 population density with 2003 UN national urban population estimates, city gazetteers, and mean urban population densities to estimate the fraction of urban land cover within a given 0.93 km pixel (Goldewijk 2001, 2005).

The urban extent measured from these maps differs from 276,000 km^2 to over 700,000 km^2 depending on remote sensing data sources and method used in these maps. In this chapter, two types of global urban land cover map created from using nighttime lights and MOIDS images are introduced.

4.4 Mapping Global Urban Land Cover from Nighttime Light Imagery

Nighttime light is a unique indicator of human activity that can be measured from space. The satellite sensor for nighttime lighting could be used to map the extent and character of development area. Since the early 1970s, the U.S. Air Force's DMSP has operated polar orbiting platforms, carrying cloud-imaging satellite sensors capable of detecting clouds using two broad spectral bands: visible–near infrared (0.4–1.1 μm) and thermal infrared with a nominal spatial resolution of 2.7 km (Elvidge et al. 2001). The operational linescan system (OLS) sensor uses a photomultiplier tube at least four orders of magnitude more sensitive than those used in any other satellite system. These images were presented as a potential urban mapping tool first by Croft (1978). The high contrast provided by this type of image and the sensor's moderate spatial resolution make it a potential choice for identifying areas where significant human activity is being carried out.

Digital DMSP/OIS datasets was prepared by NOAA's National Geophysical Data Center (NGDC). The NCDC processed DMSP-OLS data and used image time series analysis to distinguish stable lights produced by cities, towns, and industrial facilities from temporary lights arising from fires and lightning. A digital archive for the DMSP–OLS data was established in mid-1992 at NGDC. Since then, NGDC has been producing DMSP nighttime lights products and has worked extensively with the scientific community to develop applications for this data source. Figure 4.4 is an average of global nighttime light imagery in 2012. Urban areas are observed from city lights. Although there are some cultural differences in the quantity and quality of lighting in various countries, there is a remarkable level of similarity in lighting technology and lighting levels around the world (Elvidge et al. 2007b). The remote sensing of nighttime lights from urban areas provides a potential to accurate, economical, and straightforward way to estimate the global distribution and density of developed areas.

One early effort to use DMSP-OLS data to map urban area was accomplished by choosing a simple threshold technique to convert nighttime lights to urban land cover across the continental United States (Imhoff et al. 1997). A threshold of ≥89% was used to the cumulative percentage lighted data to separate dense urban core areas from most rural areas.

The global effort by using DMSP-OLS data was also conducted by using a regional impervious surface data to create a model for mapping impervious surface globally (Elvidge et al. 2007b). The model was developed using the 30 m USGS NLCD 2001 impervious surface product as the reference data source. The NLCD impervious surface that was produced from using Landsat imagery and regression tree algorithms was

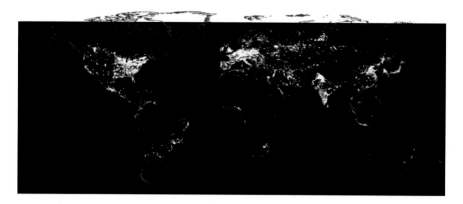

FIGURE 4.4
2012 global nighttime light image obtained from NOAA NGDC.

aggregated to a 1 km equal area grid in an Albers projection. The Landsat and radiance calibrated nighttime light images were then reprojected to the same projection in 1 km grid. Linear regression equation was created to estimate the density of impervious surface. Following restrictions were implemented:

1. Only grid cells with population count values of three or greater were included in the regression.
2. Outliers having extremely high population counts (greater than 3000 persons/km^2) and extremely bright lights (digital numbers greater than 800) were excluded in the regression.

Totally, 470,894 grid cells were included in the regression analysis. The equation developed through the regression is as follows:

$$\text{Percent impervious surface area} = 0.0795(\text{radiance})$$
$$+ 0.00868(\text{population count})$$

The regression had an r^2 of 0.59 and was highly significant $p < .0001$. The comparison between the USGS impervious surface and NGDC estimate shows that the difference is about 6699 km^2 for the conterminous United States. The NGDC estimate is slightly smaller than the estimate of the USGS NLCD product. Further analysis indicates that the two products are largely equivalent outside of the most sparsely populated areas where no lighting was detected.

The total impervious surface of the world was estimated to be 579,703 km^2. The country having the most impervious surface covers was China (87,182 km^2) followed closely by the United States (83,881 km^2), and India (81,221 km^2). Impervious surface covers of China and India were population driven, whereas the impervious surface area of the United States was more driven by affluence. The estimated global impervious surface was also used to calculate impervious surface cover per capita. The global average of impervious surface cover per capita was estimated to be 93 m^2/person.

4.5 Map Global Urban from MODIS Satellite Data

MODIS is one of instruments carried on board of the Terra platform, which was launched in December 1999. MODIS provides continuous global coverage every one to two days, and collects data from 36 spectral bands. Two bands (1–2) have a resolution of 250 m. Five bands (3–7) have a resolution of 500 m.

The remaining bands (8–36) have a resolution of 1000 m. The swath width for MODIS is 2,330 km.

The early effort to use coarse MODIS data to map urban landscape started by using a supervised classification method with MODIS data integrated with the DMSP nighttime lights dataset and population data (Schneider et al. 2003). The approach has been used to produce a 500 m global urban extent map by using a global training database and an ensemble decision-tree classification algorithm (Schneider et al. 2009, 2010).

To characterize urban land cover in the global scale, several procedures are necessary, including definition of urban extent, training data acquisition, and urban land cover classification.

4.5.1 Urban Extent Definition

The MODIS-based urban land cover mapping starts with the definition of urban extent. Comparing with moderate-resolution satellite imagery, MODIS 500 m pixel could contain a wide range of surface reflectance that belongs to different landscapes. If the VIS conceptual model is implemented with MODIS pixels, impervious surface or urban land cover usually occupies a small portion of a pixel in most urban areas. Therefore, the urban definition is important in order to identify urban extent from other land covers. The definition of urban areas is based on the physical attributes and composition of the Earth's land cover and urban areas are places dominated by the built environment. The following two constraints were used to determine urban extent (Schneider et al. 2009).

1. The coverage of the built environment including all nonvegetative, human-constructed elements, such as roads, buildings, and runways, is greater than 50% of a given landscape unit or the pixel of remotely sensed imagery.
2. Urban areas are contiguous patches of built-up land greater than 1 km^2.

These constrains define urban areas by both cover density and minimum mapping unit two major procedures, classifications of MODIS data and urban ecoregions, are required for mapping global urban land cover from MODIS imagery.

4.5.2 Classification Tool

A supervised decision tree algorithm (C4.5) was used as the classification tool for MODIS imagery classification. Decision trees have been widely used for land cover classification using remote sensing (Friedl and Brodley 1997;

Hansen et al. 1996; Pal and Mather 2003), including several studies focused on land cover characterizations from coarse-resolution data (DeFries et al. 2000; Friedl et al. 2002).

The C4.5 is a nonparametric classifier and is able to handle large, non-parametric datasets with noisy or missing data, complex, nonlinear relationships between features and classes (Friedl and Brodley 1997, Friedl et al. 2002). C4.5 is also used in the MODIS land cover product (Friedl et al. 2009). Decision tree is constructed through the recursive partitioning of a set of training data, which is split into increasingly homogeneous subsets based on statistical evaluations implemented to the feature values (the satellite imagery). Once the decision tree has been created, these rules are then applied to the entire image to produce a classified map.

The boosting technique that improves class discrimination by estimating multiple classifiers while systematically varying the training sample (Quinlan 1996) is also used in the decision tree algorithm to improve classification accuracy and efficiency. Boosting has been shown to be an equivalent form of additive logistic regression; therefore, probabilities of class membership can be assigned for each class at every pixel, resulting in an accuracy-weighted vote across all classifications.

4.5.3 MODIS Data

One-year time series of MODIS data was usually used to exploit spectral and temporal properties of land cover types. The MODIS data were processed, so that the differences in temporal signatures for urban and rural areas that result from phenological differences between vegetation inside and outside the city were intensified and utilized to segregate urban from other land covers. Procedures include the following:

1. Eight-day NBAR values for the seven MODIS land bands (463.3 m resolution) for one year
2. Monthly and yearly minima, maxima, and means for each band
3. Enhanced vegetation index (Huete et al. 2002)

All input data are adjusted to a nadir-viewing angle to reduce the effect of varying illumination and viewing geometries (Schaaf et al. 2002). The eight-day values are aggregated to 32-day averages to reduce the frequency of missing values from cloud cover.

4.5.4 Training Data

The training data were collected from sites ranging from 1 to 100 km^2 in area (Figure 4.2). These sites were characterized according to the International Geosphere–Biosphere Programme (IGBP) 17-class system (Belward and

Loveland 1997). Each site contains a polygon created by manual interpretation of Landsat and Google Earth data (4–30 m resolution) (Friedl et al. 2009), where land cover is uniform and representative of one IGBP class. An independent set of urban training sites was also chosen from 182 cities located across the globe.

4.5.5 Urban Ecoregion

Urban environments usually contain complicated heterogeneous landscapes. However, there is certain regularity in city structure, configuration, constituent elements, and vegetation types within geographic regions associated with level of economic development (Schneider and Woodcock 2008; Pickett et al. 2008). In terms of synthesis, classifications differ in how they combine social, physical, and biotic components. To facilitate processing and classification of urban land cover using the MODIS data, the Earth's land surface was stratified based on the natural, physical, and structural elements of urban areas. Here, a term of urban ecoregion was introduced to exploit these local similarities and to delineate regions that span large areas (i.e., continental scale), and that encompass multiple cities and a range of land cover types. Totally, 16 quasi-homogeneous strata were defined as urban ecoregions around the world. It is worth to note that the stratification is different from the concept of an *urban ecosystem*, where elements of the social and built environment are intermixed with biological and physical features within a given area, such as a watershed (Grimm et al. 2000; Pickett et al. 2001). Urban ecoregion focuses on large scales and contains multiple cities with different land covers. The usage of the urban ecoregion stratification with the global MODIS imagery could increase map accuracy and efficiency.

4.5.6 Classification of Global Urban Land Cover

The classification is completed by running the classification algorithm twice.

1. First classification

   The classification utilizes the full set of land cover examples that includes urban sites. The urban class probabilities are also extracted. Although the first classification focuses on characterizing the urban core and mixed urban spaces, caveats that some nonurban areas (typically shrubland) are erroneously labeled urban land may not be avoidable. The areas of potential confusion are determined based on low membership to the urban class. These confusions are removed by implementing additional classification procedure.

2. Second classification

In the second classification, the urban training sites are excluded and a posterior probabilities is calculated using Bayes' rule (Robert 1997). The probabilities from the second classification function is used as a priori information to modify the probabilities of urban land cover produced from the first. The urban probabilities are then used to determine whether a pixel should be labeled as urban land or it should be classified as another land cover class. Urban areas are typically confused with shrubland because of a relatively similar *mixed* signal of vegetation and bare ground and bright surfaces in the following urban ecoregions.

a. 4: Temperate grasslands of North–South America
b. 5: Temperate grassland in the Middle East
c. 11: Tropical–subtropical grasslands
d. 12: Temperate Mediterranean
e. 14: Arid, semi-arid steppe in Central Asia

In these regions, an assumption is made that any area having a low (or zero) probability of shrubland/savannah is likely available for urban land. The land cover probabilities from the open shrubland and savannah classes are then used to create a probability surface and served as prior information to constrain the decision tree result. The a priori probability of urban land, P(urban), is therefore estimated as

$$P(\text{urban}) = 1 - P(\text{shurbland})$$

where:

P(shrubland) is the probability of shrubland/savannah from the second decision-tree classification run in which urban training data is not included

After the probability is calculated, Bayes' rule is applied at every pixel in the region to combine the a priori information with the conditional probabilities from the decision-tree output. The posterior probabilities are then produced and are visually assessed against high-resolution data, such as aerial photos and imagery from Google Earth. An appropriate threshold is chosen to create the final map of urban extent. Here, a mean threshold of 40% is selected.

The last step is a postprocessing procedure on a region-by-region basis. A posteriori probabilities for difficult-to-classify regions are calculated. Also, regional postprocessing includes application of the MODIS 500 m water mask, and use of a spatial filter to remove single stand-alone urban pixels.

4.5.7 Accuracy Assessment for the Global-Scale Product

Accuracy assessment is accomplished by using a sample of 140 Landsat-based maps (30 m resolution) of urban areas and metropolitan regions. A random-stratified sampling design is chosen to select samples from theses cities based on population size, geographic region, and income. The sizes of these 140 cities range from 20 to 8000 km² and are independent to the training samples. For each case study city, the urban core is typically buffered by 30 km to include all potential urban areas. Moreover, random samples were drawn from high-resolution imagery using Google Earth to create independent Landsat-based reference maps for accuracy assessment.

Figure 4.5 displays continental views of MODIS 500 m global urban map in North America. High-intensity urban areas in the east coast of the United States are represented in the map. Figure 4.6 shows urban areas in Asia. Large portions of urbanized areas in eastern China and central India are shown in the map. High-intensity urban regions in Japan, Korea, and Mideast regions are also identified in the map.

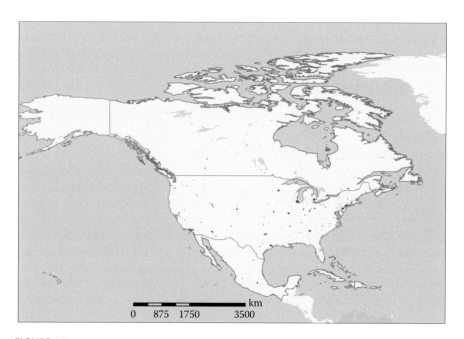

FIGURE 4.5
Continental views of the MODIS 500 m global urban map in North America. For viewing purposes, the 463.3 m resolution has been aggregated to 2 km resolution; this step yields a continuous value map where each 2 km pixel depicts the percentage of cells labeled as *urban* in the native resolution map.

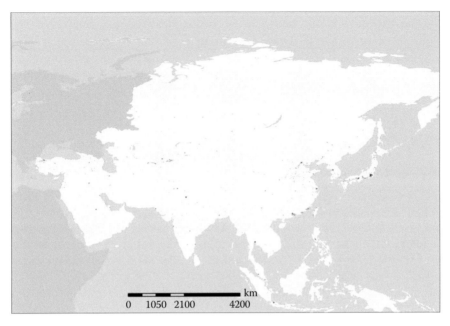

FIGURE 4.6
Continental views of the MODIS 500 m global urban map in Asia.

4.6 New Progresses on Global Urban Mapping Effort

Previous sections demonstrate different efforts to measure urban land cover globally. These works show that Earth observation sensors developed to a stage where global maps have been made possible on coarse resolution from 250 m to 2 km in the last decade. However, the coarse geometric resolution is a clear restriction, tracing the small-scale urban outlines, extents, and patterns. Also, most of these products are provided for a single time step and can be used to monitor urban land cover change.

In this section, the latest progresses on mapping global urban land cover that have been performed by Taubenböck et al. (2012) are introduced. The study focused on developing urban footprints for 27 current megacities throughout the world. These selected cities include New York, Los Angeles, and Mexico City, Mexico, in North America; São Paulo, Rio de Janeiro, and Buenos Aires in South America; Tokyo, Osaka, Seoul, Shanghai, Beijing, Guangzhou, Manila, Jakarta, Dhaka, Kolkata, Delhi, Karachi, Mumbai, Istanbul, and Tehran in Asia; Moscow, Paris, and London in Europe; and Cairo, Lagos, Kinshasa/ Brazzaville in Africa. The population development and the rate of urbanization in megacities outside the developed countries were enormous since 1975. The explosion of megacities in developing countries results from both the population growth because of the demographic transition and the process of migration because of economic gap between urban and rural productivities and incomes as well as sociological reasons. They applied both Landsat and TerraSAR-X data to create baseline urban footprints for these cities. Furthermore, historical change detections were performed using Landsat images to map urban extent changes between 1970 and 2010 for these megacities.

4.6.1 Remote Sensing Data

In this study, Landsat data in circa 1975, 1990, and 2000 are utilized. With 38 Landsat scenes per time step, all current 27 megacities are covered. All Landsat standard data products are processed using the Level 1 Product Generation System with the parameters including cubic convolution resampling method, 30 m (TM, ETM+) and 60 m (MSS) pixel size (reflective bands), universal transverse mercator map projection, WGS84 as geodetic datum, and MAP (north-up) image orientation. The standard terrain correction (Level 1T) that provides systematic radiometric and geometric accuracies by incorporating ground control points while employing a digital elevation model (DEM) for topographic accuracy is used.

The spatial extent of every image is approximately 30×50 km^2. Totally, 98 TerraSAR-X scenes between 2011 and 2012 are necessary and available to cover the spatial extents of the 27 megacities. These data have the advantage of being weather independent and consistently available for all megacities. Two un-geocoded products are available for TerraSAR-X, including the single-look slant-range complex product, which provides the slant-range imaging geometry, and the multi-look ground-range detected product. Based on characteristics of both products, classifications of urbanized areas can be derived by applying image analysis techniques to both image types. The classification results are geocoded by using the *TerraSAR multimode SAR processor* (TMSP) processing chain of the enhanced ellipsoid corrected product, which is featuring pixel location accuracies of 2 m and better depending on the accuracy of the used DEM (Huber et al. 2006). The operational SSC products using science orbits could even be lowered to 1 m. The TMSP geocoding processing chain corrects terrain-induced distortions with consideration of a DEM having a moderately coarser resolution than the TerraSAR-X products. The standard projection for enhanced ellipsoid corrected products is the same as Landsat data.

4.6.2 Image Classification

4.6.2.1 Landsat Data

Two approaches were used in the classification procedures, including an object-oriented method for the optical data and a pixel-based one for the radar data.

Generally, processing chain for classifying the multitemporal optical datasets as well as the temporally and methodologically independent classification sequence of the radar data includes three layers: data processing, object-oriented classification, and postclassification change detection. The overview of the processing chain from the multi-satellite data to the multitemporal change detection product reveals that the combination of all resampled urban footprint classification results to derive individual, but consistent and thus comparable urban footprints as basis for the change detection product.

The classification of the various Landsat scenes is based on an object-oriented classification procedure developed previously (Taubenböck 2008). The procedure consists of two main steps: multiresolution segmentation and a hierarchical thematic classification. In the first step, a bottom-up region-merging technique starts with one pixel object. In a pair wise clustering process, the weighted heterogeneity n and h of resulting image objects, where n is the size of a segment and h a parameter of heterogeneity, were minimized in the underlying optimization procedure. Adjustments of the scale parameter between 5 and 20—in dependence of the structure of the city are permitted in the multiresolution segmentation.

In the second step, a hierarchical decision tree structure was utilized to conduct classification for urban land only. For the purpose of creating fast, straightforward, and consistent classification results, the feature sets within the classification procedure, but the user can interactively adapt the threshold values. The corresponding feature selection is based on four sample Landsat scenes—three at different seasons of the year in Mexico City and one from Istanbul. These scenes cannot reflect the complete spectral diversity of urban reflectance, but they can be used as examples to reduce possible classification features. Thus, the method can be consistent when different persons are applying the algorithm to many different datasets.

After the multiresolution segmentation, the classes are further classified hierarchically, starting with classes of significant separability from other classes and ending with those of lower separability using the selected feature sets. Remaining unclassified segments not fulfilling the feature for classification are reevaluated in a further classification step. The classification of the urban footprint starts with the dataset in 2000.

4.6.2.2 TerraSAR-X Data

For the classification of TerraSAR-X data, a pixel-based classification algorithm was implemented (Esch et al. 2010). The methodology includes a specific preprocessing of the original intensity SAR data followed by an automated, threshold-based image analysis procedure.

The main purpose of the preprocessing is to provide additional texture information for classification by highlighting highly textured image regions, typically representing highly structured, heterogeneous built-up areas through taking advance of specific characteristics of urban SAR data, showing strong scattering due to double bounce effects in these areas. Therefore, the pre-processing focuses on the analysis of local speckle characteristics to provide this texture layer. In the pixel-based classification, this texture layer is used along with the original intensity information to automatically extract the urbanized areas. To analyze the local image heterogeneity C_S, the coefficient of variation is established to define the local development of speckle in SAR data as $C_S = \sigma_s/\mu_s$, where μ_s is the mean backscatter amplitude value and σ_s is the standard deviation of the backscatter amplitude in the neighborhood of the center pixel in the SAR image. Generally, the image texture C_S outcomes from a combination of the true image texture C_T and the fading texture C_F represents the heterogeneity caused by speckle as

$$C_S^2 = C_T^2 C_F^2 + C_T^2 + C_F^2$$

Urban environments or woodland usually have highly textured surfaces and these surfaces show distinct structures, leading to an increase of directional, non-Gaussian backscatter. Hence, the texture for such regions results from C_T and C_F, so that the measured image texture C_S contains comparably high values. Homogeneous surfaces without any true structuring such as grassland or noncultivated bare soil show almost no true texture C_T, suggesting that the measured image texture C_S more or less equals the fading texture C_F. Furthermore, the local true image texture C_T is defined by

$$C_T = \frac{C_S - C_F}{1 + C_F}$$

The brightest regions in the image correspond with urban areas, while a low speckle divergence characterizes open areas such as grassland or water bodies.

The speckle divergence texture file serves as input to the classification procedure that starts with the identification of potential urban scatters named as *urban seeds*. Therefore, in contrary to optical datasets, the urban seeds primarily represent the location and distribution of man-made structures with a vertical component that has strong scattering due to double bounce. Urban seeds are characterized by high amplitude in combination with a highly heterogeneous neighborhood represented by a high speckle divergence related to the true texture C_T. This was completed by comparing the locally (15×15) with regionally (45×45) averaged C_T against two thresholds. In case if one of the thresholds is exceeded, then accordingly, the image region is characterized as a distinct backscattering cluster. The urban seed layer is a binary file showing a value of 1 for seeds—indicating bright point scatters within built-up areas—and a value of 0 for all other areas.

In the second classification step, all image regions showing both a certain amount of distinct backscattering cluster in a neighborhood of 99×99 pixels and a regionally increased C_T, which is defined by two additional thresholds, are assigned as built-up area. Furthermore, distinct backscattering cluster are integrated into the class *built-up area* and all unclassified pixels are assigned as *nonbuilt-up area*. These two classes were added together to form a binary mask. The final outcome represents the urban footprint image. The characterized urban footprint contains the complicity of the urban landscape textures.

4.6.3 Multitemporal Change Detection

For the multitemporal change analysis, different spatial resolutions for urban extents should be avoided. All urban footprint classifications at all four time steps were resampled to one geometric resolution to have a consistent dataset for change detection. Products created from the MSS and the TerraSAR-X were scaled to the Landsat TM and ETM+ classifications. The resampling applied assigns the class *urban* to the higher resolution, if 75% of the pixel is covered with urbanized areas at the MSS resolution. Meanwhile, the TerraSAR-X classifications need to be downgraded. As TerraSAR-X can detect man-made vertical structures with a geometric resolution of 3 m, a 50% percentage of urbanized areas are used to define a pixel as *urbanized* at the 30 m resolution.

4.6.4 Urban Footprint and Multitemporal Change Product

The consistent dataset for the megacities of the world features four individual urban footprint classifications at four time steps in the mid-1970s, around 1990, 2000, and 2010. Two selected examples for Istanbul, Turkey (a) and Mexico City, Mexico (b), reveal urban land cover extents and development intensities in these regions. Generally, a number of trends can be immediately observed in terms of spatial extent and the spatial dynamics of urbanization over the time. The accuracy assessment also showed on average that the overall accuracy came up to 90.5% with a variance in the assessment of 12.5. For these two specific areas, overall accuracies are 88.0%, 93.0%, 92.0%, and 88.4% for Landsat MSS, Landsat TM, Landsat ETM+, and TerraSAR-X data, respectively, derived urban footprint in Mexico City. The overall accuracies for Istanbul are 93.7%, 92.4%, 90.8%, and 92.8%, respectively, for the same datasets.

4.7 Summary

Understanding the distribution and dynamics of the world's urban land cover is essential to better understand the Earth's fundamental characteristics and processes, including biogeochemical cycles, hydrological cycles, and biodiversity in built-up lands. Current efforts for global-scale urban land map are based on coarse-resolution satellite imagery. The coarse-scale urban map can provide relatively quick and reliable estimate for global urban land distributions. However, urban land does not always occupy a large piece of land that can be detected from coarse-resolution remotely sensed imagery. This chapter describes satellite-based global land cover datasets at both national and global levels and principles to produce these products. These urban land cover datasets were created using different classification methods. Urban land cover dataset in the United States has been characterized from percent impervious surface in three time periods. This dataset captures spatial distribution patterns and development intensity in details because the subpixel level impervious surface product is characterized and used to define urban land cover. The urban land cover characterized from MODIS imagery provides relatively accurate map about global distribution of built-up land. However, the coarse resolution of MODIS imagery limits the product to capture some details of fundamental features in urban areas, especially for monitoring new urban growths in medium- to low-intensity developments. The availability of high-resolution Landsat 30 m satellite data and advancement in remote sensing and information technology, improvement in urban land cover mapping and monitoring methodologies, and increasing availability of resources and expertise in the last decade have provided an opportunity for producing urban land cover dataset of the world at higher spatial and temporal resolutions. Systematic large area urban land cover monitoring at 30 m resolution is possible by using urban land cover information in such scale to meet varieties of global application.

5

Assessment of Water Quality in Urban Areas

5.1 Introduction

Urbanization brings substantial changes to the Earth's surface in the physical and biophysical environments. One of the major human activities that has an impact on the environment is quality of inland water environments. Direct environmental impacts of increasing urban land use include the degradation of water resources and water quality from either point or non–point sources. Point sources can be traced to a single source, such as a pipe or a ditch. Non–point sources are distributed when surface runoff transports non–point source pollutants from their source areas to receiving lakes and streams (Gove et al. 2001; USEPA 2001; Peng 2012) and are associated with the landscape and its response to water movement, land use and management, and/or other human and natural activities on the watershed. Agriculture, industrial, and urban areas are anthropogenic sources of point and nonpoint substances. Pollutants either dissolved or suspended in water or associated with sediment, including nutrients, heavy metals, and oil and grease, can accumulate and wash away from impervious surface areas, and lead to deterioration of water quality, which affects most freshwater and estuarine ecosystems in the world.

Frequently, impervious surface has been considered as a key environmental indicator of the health of urban watersheds (Schueler 1994) and as an indicator of non–point source pollution or polluted runoff (Arnold and Gibbons 1996; Slonecker et al. 2001; Xian et al. 2007). With an increase in the demands from water sources for many cities, there is increasing interest in protecting the qualities of existing water resources and for making more use of the total available water resources to urban areas. To better understand the underlying processes and develop adequate policies for environmental management, restoration, and adaptation, and eventually reduce the negative impacts of these changes, changes in water quality associated with urbanization need to be accurately detected and identified.

Most water quality issues for urban population usage are associated with urban development, industrial water usage, and agricultural activities. Lakes, rivers, and reservoirs inside or nearby urban areas are usually facing a number of serious challenges related to water resource management, which include population growth, water scarcity, water resource degradation, and water pollution (Cavalli et al. 2009; Oki and Yasuoka 2008; Volpe et al. 2011). For example, human-made reservoir water is the main source of drinking water and industrial and agricultural water usages in many urban areas. Most reservoirs accumulate the runoff formed of rainwater, and some reservoirs accumulate the water from rivers with long distance drainage. Local industrial developments and population increases have affected reservoir ecosystem and water quality in many urbanized watersheds. Figure 5.1 shows water conditions in 2001 and 2004 in Dai Lake, Kunming, China. The city of Kunming located on the north side of the lake is inhabited by more than five million people. Urban growth around the Kunming area has deteriorated the water quality of the lake and the surrounding water bodies. Satellite observations captured the change in water quality of the lake and surrounding water bodies.

As water quality variations related to increasing in human population and urbanization pressures continue in coastal and inland areas (Caraco 1995; World Resources Institute 2003), effective water quality monitoring and assessment have become critical for water resource management and sustainable development. Without accurate, intensive, and long-term water quality data acquisition, the state of the world's water resources cannot be adequately assessed.

Traditionally, water quality evaluation has depended on costly, time- and labor-intensive on-site sampling and data collection. These research and

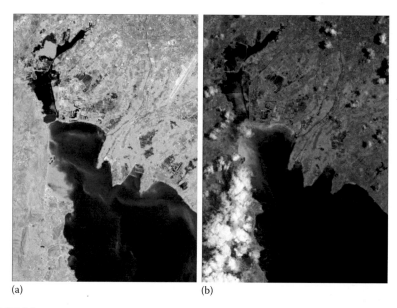

(a)   (b)

FIGURE 5.1
Landsat images acquired in (a) June 2001, and (b) June 2004. The water body is Dai Lake, Kunming, China.

monitoring efforts have been too limited on temporal and spatial scales to address factors that can impact development of events such as harmful algal blooms, oxygen depletion, and contamination of inland water. They do not give either the spatial or the temporal view of water quality needed for the accurate assessment or management of water bodies.

On the other hand, substances in surface water can significantly change the backscattering characteristics of water. Remote sensing techniques provide the ability to measure these spectral features through the spectral signature backscattered from water. The measured features can then be related to a water quality parameter by empirical or analytical models (Ritchie et al. 2003; Matthews 2011). The optimal wavelength used to measure a water quality parameter is dependent on the substance being measured, its concentration, and the sensor characteristics.

The factors that affect water quality are complicated and vary among water bodies. Major components include suspended sediments (turbidity), algae (i.e., chlorophylls and carotenoids), chemicals (i.e., nutrients, pesticides, and metals), dissolved organic matter (DOM), thermal releases, aquatic vascular plants, pathogens, and oils. Components such as suspended sediments, algae, DOM, oils, aquatic vascular plants, and thermal releases can alter the energy spectra of reflected solar and/or emitting thermal radiation from water bodies and can be measured using remote sensing techniques. However, most chemicals and pathogens do not directly change the spectral or thermal properties of surface water and they can only be inferred indirectly from measurements of other water quality parameters that are affected by and related to these chemicals.

Remote sensing techniques provide synoptic view and temporal coverage of surface water quality parameters that are usually not available from *in situ* measurements and therefore making it possible to monitor the surface water effectively and efficiently by identifying and quantifying water quality parameters. Passive remote sensors on many medium-resolution satellites can measure the light in the visible and near-infrared (NIR) part of the electromagnetic spectrum (400–1000 nm) and are most often used for water-related applications. The optically active water constituents, including phytoplankton, tripton made up of detritus and minerals, colored DOM (CDOM), and water itself, all can influence the optical signature of water in the visible wavelengths. Figure 5.2 illustrates that the radiance leaving surface water is modified through the backscattering and absorption of light by these constituents (Matthews et al. 2010). Water shows strong absorption at wavelengths >750nm and therefore signals from other constituents except in highly turbid water, where scattering by minerals overwhelms absorption by water, could not be detected above that spectral range. Generally, wavelengths between 400 and 750 nm contain the most information on the water constituents, which is detectable by remote sensing instruments, with the exception of highly turbid water where the signal in the NIR is also useful. Moreover, the historical archive of global satellite observations and recent launch of many new satellite instruments have provided capabilities for

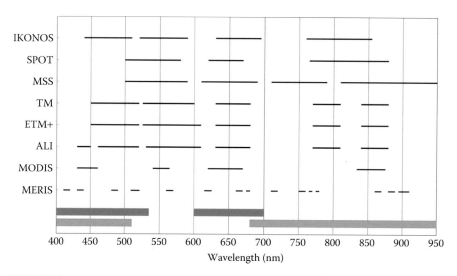

FIGURE 5.2

The spectral range of various satellite instruments in relation to the location of the maximum influence of absorption by phytoplankton, a_ϕ, detritus and gelbstoff, a_{dg}, and water, a_w. The bands plotted for MODIS are the 250 and 500 m bands. (Modified from Matthews, M.W., *Int. J. Remote Sens.*, 32, 6855–6899, 2011.)

successful water-related remote sensing applications and improved real-time monitoring of water quality and the rapid detection of environmental threats in urban areas. Table 5.1 illustrates the most frequently used satellite sensors and their spectral bands and spatial and temporal resolutions for the detection of water quality parameters in inland and transitional water bodies.

In this chapter, different methods used to detect different constituents in surface water by different satellite sensors are introduced. Section 5.2 describes the most commonly used parameters for qualifying water quality. Section 5.3 illustrates the use of remote sensing data to derive several parameters for water quality assessment. Section 5.4 introduces the evaluation of nonsource pollutant loadings in a watershed. Section 5.5 gives several examples of using satellite remote sensing data to assess water quality. Section 5.6 is the summary of the chapter.

5.2 Parameters of Inland Water Quality

Generally, the methods used for analyzing water quality with remote sensing data can be identified into two types: empirical and semi-analytical. Empirical methods use *in situ* data on the characteristic of interest to calibrate a predetermined fixed model or regression analysis to determine the *best-fit* model using radiance data from available bands and band combinations. Semi-analytical

TABLE 5.1

Satellite Sensors and their Spectral Band(s) and Spatial and Temporal Resolutions That Are Most Often Used for the Detection of Water Quality Parameters in Inland and Transitional Water

Satellite	Sensor	Spectral Resolution (μm)	Spatial Resolution	Temporal Resolution
LM900	IKONOS	0.45–0.85 (4 bands)	4 m	3/5 days
IRS-P6	LISS4	0.52–0.68 (3 bands)	5.8 in	5 days
SPOT 5	HRG	0.48–1.75 (5 bands)	10 m	26 days
Proba-1	CHRIS	0.415–1.050 (19 bands)	18 m	~7 days
SPOT 4	HRVIR	0.50–0.89 (3 bands)	20 m	26 days
IRS-P6	LISS3	0.52–1.70 (4 bands)	23.5 m	24 days
EnMAP	HIS	0.420–2.450 (200 bands)	30 m	4 days
EO-1	*Hyperion*	0.4–2.5 (220 bands)	30 m	16 days
EO-1	ALI	0.43–2.35 (9 bands)	30 m	16 days
Land sat 5	TM	0.45–2.35 (6 bands)	30 m	16 days
Land sat 7	ETM+	0.45–2.35 (8 bands)	30 m	16 days
1SS	HICO MOD	0.3–1.0 (128 bands)	100 m	–
Terra/Aqua	IS	0.620–0.876 (2 bands)	250 m	1–2 days
EnviSAT	MERIS	0.412–0.900 (15 bands)	~300 m	2–3 days
Sentinal 3	OLC	0.413–1.020 (16 bands)	~300 m	2–3 days
1RS-P4	WiFS	0.402–0.885 (8 bands)	1 km	2 days
COMS	OCM	0.400–0.885 (8 bands)	360 m	2 days
TRAQ	GOCI	0.400–0.900 (8 bands)	500 m	~15 min
Sea WiFS	OCAPI	0.320–2.13 (8 bands)	4 km	14 min

Source: Matthews, M.W. *Int. J. Remote Sens.*, 32, 6855–6899, 2011, Table 2.

methods use models of radiative transfer through the atmosphere and water. These models usually require accurate atmospheric correction and radiation field measurements of inherent optical properties for the entire range of conditions in the water bodies being assessed (Olmanson et al. 2011). Most current studies suggest that the application of empirically based algorithms can produce common biogeophysical products for inland and transitional water bodies. For example, the band 709/664 ratio is remarkably effective for detecting high-biomass water bodies typical of eutrophic/hypertrophic systems.

Parameters are often used to create empirical relationship to measure surface water quality. Those that can be derived from remotely sensed data include phytoplankton pigments such as chlorophyll-a (Chl-a) and algal (Yacobi et al. 1995; Brivio et al. 2001; Giardino et al. 2005; Gitelson et al. 2009), cyanobacterial pigment phycocyanin (PC) (Ruiz-Verdú et al. 2008), concentration of total suspended solids (TSS) (Onderka, and Pekarova 2008; Doxaran et al. 2009; Oyama et al. 2009), absorption by CDOM (Kutser et al. 2005), Secchi disk depth (ZSD) or water clarity (Olmanson et al. 2008), turbidity, and water temperature (Matthews 2011).

Phytoplankton pigment concentration (Chl-a) is the most commonly derived parameter in water-quality remote sensing mainly because of its use in determining the trophic status of water bodies. It acts as a proxy for phytoplankton concentration and is therefore an important component in the derivation of secondary products such as primary production. Most phytoplanktons are too small to be individually seen with the unaided eye. However, when present in high enough numbers, they may appear as a green discoloration of the water due to the presence of chlorophyll within their cells. Figure 5.3 shows variations of spectral reflectance associated with the Chl-a concentration. Generally, the detection of Chl-a requires that the remote sensing reflectance be resolved in sufficient detail for the application of suitable detection algorithms. The ground resolution of a pixel for larger inland water bodies ($>\sim 1$ km^2) should be a few hundred meters or less, several times larger than the dimensions of the target.

CDOM is the optically measurable component of the DOM in water. Also known as *chromophoric DOM*, yellow substance, and gelbstoff, CDOM occurs naturally in aquatic environments primarily as a result of tannins released from decaying detritus. CDOM most strongly absorbs light of short wavelength ranging from blue to ultraviolet, whereas pure water absorbs longer wavelength red light. Therefore, nonturbid water with little or no CDOM

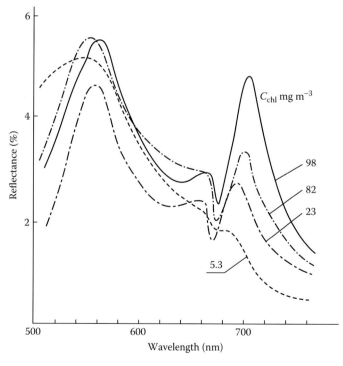

FIGURE 5.3
Spectral reflectance as measured in Lake Balaton in September 1985 with Chl-a concentrations as shown. (From Gitelson, A. et al., *Int. J. Remote Sens.*, 14, 1269–1295, 1993, Figure 1.)

appears blue. The color of water will range between green, yellow-green, and brown as CDOM increases.

TSS, which are measured per volume of water including inorganic (minerals) and organic (detritus and phytoplankton) components, are the most common pollutant both in weight and volume in surface water of freshwater systems. TSS increase the radiance emergent from surface water (Figure 5.4) in both visible and NIR proportions of the electromagnetic spectrum (Ritchie et al. 1976). TSS is important for water quality management, because it is related generally to primary production, sediment transport, and, more specifically, water clarity/opacity, which is an indicator of water quality (Dekker et al. 2002). Significant relationships between suspended sediments and reflectance from spectral wave bands suggest that reflectance of optimum wavelength can be used to estimate TSS concentration by empirical algorithms.

Other parameters are also used to measure water quality such as thermal pollution that exists when biological activities are affected by changing the temperature of a water body by anthropogenic activity. Remotely sensed data have been used to map thermal discharge into streams, lakes, and coastal water bodies from electrical power plants (Gibbons et al. 1989). Thermal plumes in river and coastal water bodies can be accurately estimated by remote sensing techniques. Mapping of absolute temperatures by remote sensing provides spatial and temporal patterns of thermal releases, which is useful for managing thermal releases.

Table 5.2 lists band(s), band ratios, and/or band arithmetic for the detection of water quality parameters in inland and transitional water bodies

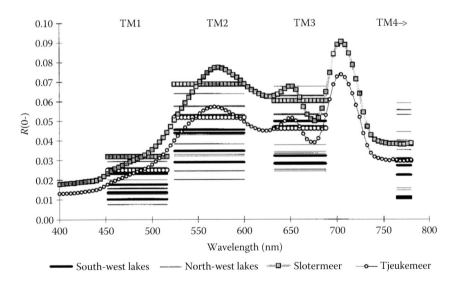

FIGURE 5.4
The Rs spectra of the Frisian water, the Netherlands, from *in situ* measurements and Landsat TM bands 1 to 4 with the full spectrum of the two largest lakes. (From Dekker, A.G. et al., *Int. J. Remote Sens.*, 23, 15–35, 2002, Figure 8.)

TABLE 5.2

Different Bands, Band Ratios, and Algorithms for the Detection of Water Quality Parameters in Inland and Transition Water

	Sensor Spectral Resolution		
Parameter	Broad Bands	Narrow Bands	Bio-Optical Basis
ZSD	Red band or red/ blue ratio, for example, TM3/ TM1 + TM1	Red band or blue/red ratio, for example, 512/620	Reflectance in red $\sim b_{bp}$–the blue band dominated by a_ϕ and aCDOM serves to normalize
TSS	<10 g m^{-3}: Red/ green ratio or (green + red)/2	<30 g m^{-3}: (560 − 520)/ (560 + 520) or single red band, for example, 700	The a_ϕ minimum at 560 nm is sensitive to TSS, whereas the 520 nm band serves to normalize
	>10 g m^{-3}: Red or NIR band or (green + red)/2	>30 g m^{-3}: NIR ratio, for example, 850/550	Reflectance in red and NIR $\sim b_{bp}$ and b_{bm}. Band ratios normalize for variations in particle refractive indices and grain sizes
	>30 g m^{-3}: NIR/ red or NIR/ green ratio		
Chl-a	<20 mg m^{-3}: Green/blue ratio or (blue − red)/ green	<30 mg m^{-3}: 560 or fluorescence line height algorithm	Chl-a $\sim$ reflectance in red due to b_{bp}, and inversely related to reflectance in blue due to a_ϕ
	>20 mg m^{-3}: Red/blue or red/NIR ratio	>30 mg m^{-3}: 700/670 ratio or three-band model $750(1/670 − 1/710)$ or RLH or SUM algorithms	Reflectance at 700 nm sensitive to $b_{b\phi}$ normalized by the a_ϕ maximum near 665 nm
aCDOM	Green/red ratio	Red/blue ratio, for example, 670/412, or *decoding index* $[490 − (700/675) − 520]/$ $[490 + (700/675) + 520]$	Relatively insensitive sensors: Reflectance in green inversely related to aCDOM normalized by reflectance in red
			Sensitive sensors: Reflectance in blue inversely related to aCDOM normalized by the reflectance in red
Tubidity	Red band	Red or NIR band	Reflectance in red and NIR $\sim b_{bp}$ and b_{bm}
SPIM	Red or NIR band	Red or NIR band	Reflectance in red and NIR $\sim b_{bp}$ and b_{bm}
PC	–	(620/650) or (709/620) ratio or $(600 + 648)/2 − 624$	Reflectance at 620 nm inversely related to PC due to absorption maximum

Source: Matthews, M.W. *Int. J. Remote Sens.*, 32, 6855–6899, 2011, Table 3.

using broadband or narrow-band sensors. In this chapter, only methods used to estimate concentration of Chl-a, CDOM, and TSS from remote sensing data are introduced.

5.3 Satellites Measurements and Empirical Algorithms for Different Water Quality Parameters

5.3.1 Chl-a and Algal

Monitoring the concentrations of Chl-a and algal/phytoplankton is important for managing eutrophication in inland water bodies. Remote sensing has been used to characterize chlorophyll concentrations spatially and temporally. As with suspended sediment measurements, most remote sensing studies of chlorophyll in water are based on empirical relationships between radiance/reflectance in narrow bands or band ratios and chlorophyll. The correlation between Chl-a concentration and the ratio is very significant ($r^2 > 0.8$) for a variety of water bodies, including rivers, lakes, estuaries, and in the laboratory, and over a wide range of concentrations from ~0.1 to 350 mg m^{-3} (Moses et al. 2009).

Suspended sediment measurements show that most remote sensing studies of chlorophyll in water are based on empirical relationships between radiance/reflectance in narrow bands or band ratios and chlorophyll. Spectra with increasing reflectance also show increasing in chlorophyll concentration across most wavelengths but areas of decreased reflectance in the spectral absorption region for chlorophyll (675 to 680 nm) (Schalles et al., 1997). Figure 5.4 illustrates that backscattering from particulate matter (phytoplankton) and the strong absorption of water both increase toward the infrared. The offset to scattering due to absorption by water near 700 nm results in a sharp peak in highly scattering (turbid or productive) water bodies. The height and position of the peak is known to be well correlated with Chl-a, with the peak shifting toward greater wavelengths of 715 nm as Chl-a increases (Gitelson 1992; Schalles et al. 1997). In contrast, the reflectance near 670 nm is uncorrelated with Chl-a. Thus, the 700/670 nm ratio can be effectively exploited to determine Chl-a, as it normalizes the signal from particulate phytoplankton backscattering. The ratio of reflectance at about 700 nm to that near 670 nm has been widely used for estimating Chl-a concentration in high-biomass water bodies.

Generally, empirical relationships between spectral properties and water quality parameters can be established. The general forms of these empirical equations are based on either linear or nonlinear regression analysis. These equations also are dependent on satellite sensor.

A logical linear equation has been used based on aircraft measurements to determine seasonal patterns of chlorophyll content in the Chesapeake Bay, Maryland (Harding et al. 1995).

$$\text{Log}_{10}(\text{chlorophyll}) = a + b(-\log_{10}G) \tag{5.1}$$

where:
  a and b are empirical constant derived from *in situ* measurements
  G is $R_2^2/(R_1 \times R_3)$
  R_1 is radiance at 460 nm, R_2 is radiance at 490 nm, and R_3 is radiance at 520 nm

This algorithm was used to map total chlorophyll content in the Chesapeake Bay.

A three-band model that focused on variation on the band ratio (Gitelson et al. 2003) can give a good estimate of Chl-a in turbid and very high biomass hypertrophic water bodies. The model is defined as

$$\text{Chl-a} = R(\lambda_3) \times \left[\frac{1}{R(\lambda_1)} - \frac{1}{R(\lambda_2)} \right] \tag{5.2}$$

where:
  Chl-a is in unit of mg m^{-3}
  $R(\lambda_1)$ is reflectance in wavelength maximally sensitive to Chl-a absorption (670 nm)
  $R(\lambda_2)$ is reflectance in wavelength minimally sensitive to absorption by Chl-a (710 nm)
  $R(\lambda_3)$ is reflectance in wavelength minimally effected by absorption that accounts for scattering (750 nm)

A more complicated algorithm (Matthews et al. 2012) can be used to detect trophic status (chlorophyll-a), cyanobacteria blooms, surface scum, and floating vegetation in coastal and inland water bodies. The algorithm uses top-of-atmosphere data from the medium-resolution imaging spectrometer (MERIS) in red and NIR bands to avoid error-prone aerosol atmospheric correction procedures used to estimate the reflectance from surface water, while the baseline subtraction between red and NIR bands minimizes the atmospheric effects from the aerosol particles. The algorithm connects Chl-a concentration with the maximum peak-height (MPH) of articulate backscatter/absorption in the MERIS red/NIR bands through regression analysis. MPH is calculated using a baseline subtraction algorithm to calculate the height of the dominant peak across the red and NIR MERIS bands between 664 and 885 nm caused by sun-induced chlorophyll fluorescence and particulate backscatter.

The MPH algorithm, which is similar in form to the fluorescence line height algorithm (Gower et al. 1999), explores both position and magnitude of the maximum peak in the red/NIR MERIS bands at 681, 709, and 753 nm (bands 8, 9, and 10). A baseline between MERIS bands 7 (664 nm) and 14 (885 nm) is used to measure the height of the red peak and this baseline was found to give more robust results than a spectrally shifting baseline. The MPH variable is calculated as follows:

$$MPH = R_{max} - R_{664} - \left[\frac{(R_{885} - R_{664})(\lambda_{max} - 664)}{(885 - 664)} \right] \qquad (5.3)$$

where:

R_{max} and λ_{max} are the magnitude and position of the highest value, respectively, from MERIS bands at 681, 709, and 753 nm

The MPH algorithm is designed to deal with phytoplankton-dominant and harmful algal bloom- affected water. Therefore, Chl-a can be estimated by using separate fits to best estimate water qualities in different conditions.

- *The fluorescence 681 nm domain (mixed oligotrophic/mesotrophic low-medium biomass water bodies)*

$$Chl\text{-}a = 2.72 + 6903.13 \times MPH \qquad (5.4)$$

Relatively small mean absolute percentage error (69%) and high correlation ($r^2 = 0.71$) have been found for Chl-a in the range of 0.5–30 mg m^{-3}. Besides, the algorithm is sensitive to a minimum Chl-a value of approximately 3.5 mg m^{-3}.

- *The backscatter/absorption 709 nm domain*

It was found that there was a large difference in MPH values between data points from *Microcystis* cyano-dominant water (Zeekoevlei and Hartbeespoort) and those from eukaryote-dominant water (Loskop/ Benguela). Meanwhile, there was significant overlap between the data for each of these water types. Therefore, separate fits are needed for *Microcystis* cyano-dominant water (prokaryotes) and for water with phytoplankton assemblages made up predominately of dino-flagellates or diatoms (eukaryotes). For dinoflagellate- or diatom-dominant water, the best fit is expressed as

$$Chl\text{-}a = 37.18 + 11228.38 \times MPH \qquad (5.5)$$

This algorithm was found to have relatively low statistical significance and a large mean absolute percentage error. This relatively poor performance might be caused by the small sample size and the large range of Chl-a values over which the algorithm is expected to perform (a range of 343 m gm^{-3}). A modified model can be used for *Microcystis* cyano-dominant water:

$$Chl\text{-}a = 22.44 \exp(35.79 \times MPH) \qquad (5.6)$$

The nonlinear least squares estimation improves modeling performance by reducing the mean absolute percentage error and increasing correlation. However, the algorithm performs well only for Chl-a values greater than 22.4 mg m^{-3}. Furthermore, a single continuous algorithm is used for Chl-a estimation in eukaryote-dominant water. The algorithm is developed by a fourth order polynomial and can be implemented for operational use. The equation is defined as

$$\text{Chl-a}(\text{eukaryotes}) = (5.24 \times 10^9 \text{MPH}^4) - (1.95 \times 10^8 \text{MPH}^3)$$
$$+ (2.46 \times 10^6 \text{MPH}^2) + (4.02 \times 10^3 \text{MPH}) + 1.97$$

(5.7)

The mean absolute percentage error between the derived function and the unsorted data is 59.9% and the r^2 value is 0.71. The operational algorithm uses the maximum value of several band ratios, similar to the maximum peak selection of the MPH algorithm. The polynomial fit provides good continuity between the different domains of the algorithm and to obtain the smallest difference between predicted and observed Chl-a.

5.3.2 Colored DOM

CDOM is composed of humic and fulvic acids and is also named as *yellow substances* that significantly contribute to water color due to strong absorption in the blue region of the spectrum by humic substances and turning the water brown. Absorption by CDOM usually occurs at 440 nm and proceeds as a form of an exponential curve decreasing toward longer wavelengths, so that its effects are usually negligible at wavelengths >550 nm. The slope of the curve is mostly predictable as it varies within a relatively small range (0.10–0.20 nm^{-1}) for most inland and coastal water bodies (Dekker 1993). Absorption by other materials such as detrital material and mineral and nonpigmented aquatic particles exhibits similar decreasing exponential functions (Babin and Stramski 2002; Babin et al. 2003) and can overwhelm the contribution to absorption by CDOM in natural water if concentrations are high enough.

The retrieval of absorption by CDOM (aCDOM) using remote sensing is therefore of great interest from a bio-optical perspective. Because the signal from aCDOM is only significant in the blue region of the spectrum (<550 nm), it is reasonable to propose the retrieval from remote sensing by utilizing suitable bands from this region of the spectrum. However, atmospheric scattering, which is the greatest in the blue, could potentially diminish the water-leaving signal from strong absorption by phytoplankton and CDOM, so that the signal from the water may be indistinguishable and the data unusable. This limitation is more evident when using the low radiometric sensitivities of sensors such as Landsat or SPOT. Despite these limitations, algorithms have been developed for aCDOM retrieval using ratios of reflectance in the blue (~400–500 nm) to that in the green or red (~500–700 nm) so that aCDOM could be well correlated (Kutser et al. 1998). To retrieve aCDOM from high-resolution radiometric data, a reflectance ratio algorithm can be used to relate reflectance data to aCDOM (Gitelson et al. 1993).

$$aCDOM = a \left[\frac{(R_{480} - R_{700})/(R_{675} - R_{520})}{(R_{480} + R_{700})/(R_{675} + R_{520})} \right]^b \quad (5.8)$$

where:
   a and b are regression coefficients

The 480 nm band is strongly influenced by aCDOM, whereas the band at 520 nm is a reference and the R_{700}/R_{675} ratio a correction factor for Chl-a absorption. The nonlinear power-law algorithm gave an r^2 value of greater than 0.9 for aCDOM at 380 nm in the range 0.1–12 m^{-1} in >20 inland water bodies.

Broadband terrestrial sensors can also be used successfully to estimate aCDOM in lakes and transitional water, although usually not with the same level of accuracy as with sensitive and high spectral definition instruments.

5.3.3 TSS Concentration Estimate

Early study using *in situ* observations indicated that wavelengths between 700 and 800 nm were most useful for determining suspended sediments in surface water (Ritchie et al. 1976). It is apparent that empirical algorithms used to estimate concentration of TSS should focus on these spectral ranges. Because the organic and inorganic material consisting of the suspended matter has different optical properties, an average ration of organic to inorganic is usually necessary to determined for different water bodies (Dekker et al. 2001). For example, a difference ratio algorithm of the form $(R_{560} - R_{520})/(R_{560} + R_{520})$ is highly correlated with TSS in lakes and rivers not exceeding 66 g m^{-3} (Gitelson et al. 1993). The algorithm takes advantage of the phytoplankton absorption minimum near 560 nm, where the reflectance is sensitive to changes in TSS, while the reflectance at 520 nm is relatively insensitive to changes in TSS. Thus, the difference ratio acts to normalize the signal at 560 nm for scattering by TSS.

To use Landsat Thematic Mapper (TM) data for turbid water bodies, the reflectance model that uses two Landsat TM bands to estimate subsurface irradiance is appropriate (Dekker et al. 2001, 2002).

$$Rs = \frac{rR_b}{a_b + R_b} \quad (5.9)$$

where:
   r accounts for the anisotropy of the downwelling light field and the scattering processes in the water
   a is the absorption coefficient (m^{-1}) for band b

Rs is the subsurface irradiance reflectance

R_b is the scattering coefficient for band b. Landsat TM band 1 has a low
    Rs signal, with a range of 0.5%–3% because of too many substances
    absorb significantly at these wavelengths

TM bands 2 and 3 have a similar response, because all variable spectral information within a broad TM band is averaged to one Rs value. For Landsat TM band, 4 Rs has a linear relationship with increasing TSS and would be suitable for estimating TSS. However, Landsat TM has a lower sensitivity for band 4 and this band is less reliable in absolute calibration. Therefore, Landsat TM bands 2 and 3 are most suited for estimating TSS in turbid water bodies. Ultimately, an exponential relationship can be created between TSS concentration and Rs derived from bands 2 and 3.

5.4 Characterization of Pollutant Loading in a Watershed

Rapid growth of built-up land can affect surface water quality when total runoff increases, and hydrologic impairment leads to erosion and sedimentation. It is important to characterize the locations of sources of degraded stormwater runoff quality within drainage basins and identify areas that are hazardous to the beneficial uses of receiving water. Using land cover information derived from satellite remote sensing information in combination with other spatial analysis models for predicting surface pollutant loadings associated with urban development has been demonstrated to be an effective approach (Xian et al. 2007). In this study, the effects of urbanization processes on water quality and environmental conditions in the Tampa Bay watershed have been illustrated.

Tampa Bay, located on the Gulf Coast of west-central Florida, has an areal extent of approximately 1030 km^2 and is one of the largest open-water estuaries in the southeastern United States. The Tampa Bay watershed has a spatial extent of approximately 6600 km^2 and encompasses most of Pinellas, Hillsborough, and Manatee Counties and portions of Pasco, Polk, and Sarasota Counties. Four major sources of surface water—the Hillsborough, Alafia, Little Manatee, and Manatee Rivers—flow into the bay. The largest municipalities within the watershed are Tampa, St. Petersburg, Clearwater, and Bradenton. More than two million people resided in the watershed by 2000. Recent urban land use development has extended to the northeastern side of Tampa, where large open lands were available (Xian and Crane 2005).

Anthropogenic inputs of phosphorus can be large for a watershed where a large number of humans, domestic animals, and livestock coexist. To evaluate water quality in a watershed, land cover/use and nonpoint pollutant source data are required to conduct causal relationship analysis. To measure the spatial extent of urban land cover and evaluate its environmental influence without losing urban land cover and land use heterogeneity, subpixel percent impervious surface was chosen as an indicator for identifying both spatial extent and intensity of urbanization in the watershed. Subpixel percent imperviousness was estimated using the approach illustrated in Chapter 3. Both high-resolution orthoimagery and Landsat imagery were used to extrapolate impervious surface through regression models for the entire watershed. Nonpoint pollutant loadings were provided in a geographic information system format by the Engineering Division, Hillsborough County (Hillsborough County 1999) and were used to investigate the spatial variation of water quality parameters. The pollutant loading data collected for 16 subdrainage basins within Hillsborough County were used to estimate pollutant loading by using the Hillsborough County Pollutant Loading and Removal Model (PLRM) for the watershed within the county. The PLRM was developed by the County's public works/stormwater management environmental team and was implemented as the pollutant loading assessment tool for the county. The model involves calculation of gross pollutant loads, estimation of net loads based on existing treatment, and evaluation of water quality level of service based on a comparison of existing loads to the single family residential benchmark. Several water quality parameters that represent domestic (organic), as well as industrial nonpoint pollutants, including five-day biological oxygen demand (BOD5), TSS, total Kjeldahl nitrogen (TKN), total nitrites and nitrates ($NO_3 + NO_2$), total nitrogen (TN), total dissolved phosphorus (TDP), oil and grease, lead (Pb), and zinc (Zn), were evaluated. Generally, the Hillsborough River basin has the smallest annual loading for all selected sources, whereas most of the annual loadings are highest in the Alafia River basin. Inspection of land use information in the region showed that most BOD5 and nitrogen loadings in Hillsborough County are concentrated in agricultural and residential areas. Intensive landscape maintenance in residential neighborhoods also increases TKN and TDP values. However, the TSS value (Hillsborough County 1999) is lower than the average of the United States and reflects less soil erosion and effective regulations for construction. Lead data for the county are relatively lower than for other locations in Florida and this may indicate decreased emissions due to use of unleaded gasoline. Figure 5.5 illustrates spatial distributions of these nine different pollutant loadings in the watershed. Among the amounts of nine non–point source pollutant loadings, the largest pollutant value is for TSS at 156.52 ton/km^2/year. BOD5 has the second largest loading amount of 36.73 ton/km^2/year. TN also shows a relatively large amount of 11.53 ton/km^2/year loaded to the watershed. Pb and Zn metals have similar loading values. High loadings for TSS, and oil and grease are observed in the northwest corner of the watershed where population densities are relatively high. High TSS also follows major highways in the

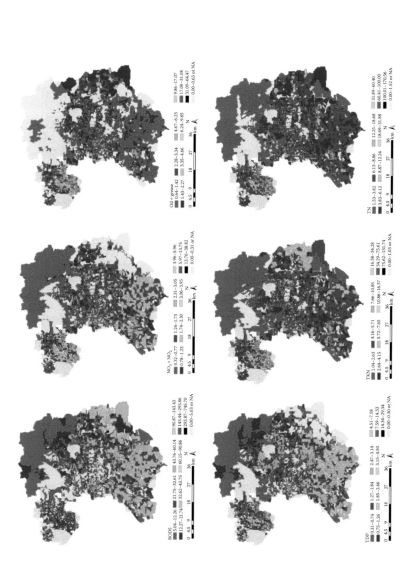

FIGURE 5.5

Annual BOD5, $NO_3 + NO_2$, oil and grease, TDP, TKN, TN, TSS, Pb, and Zn loading (ton/year) in the 16 subdrainage basins, Hillsborough County, Florida.

(Continued)

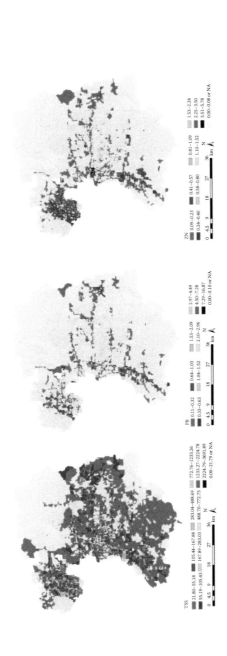

FIGURE 5.5 (Continued)

Annual BOD5, $NO_3 + NO_2$, oil and grease, TDP, TKN, TN, TSS, Pb, and Zn loading (ton/year) in the 16 subdrainage basins, Hillsborough County, Florida.

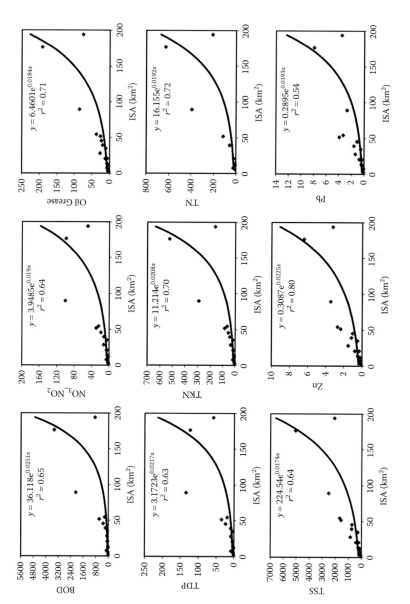

FIGURE 5.6
Polynomial regressions of nine non-point source pollutant loadings and associated ISA in each sub-drainage basins. ISA is in km² and pollutant parameters are in tons/year.

area. The spatial distribution patterns of TN and TKN are similar and both loadings have larger values in these basins that have a large portion of built-up land. The magnitude of $NO_3 + NO_2$ is relatively larger in the northwest corner and southern part of the watershed, where a greater amount of residential activity is observed from impervious surface area (ISA) distribution. Loadings of metals (Pb and Zn) are seen along major transportation roads, especially on the eastern coast of the bay. All major nonpoint pollutant sources exhibit strong regional characteristics from their spatial distribution patterns in the watershed.

The ISA product was used to extrapolate the spatial extent of the built-up areas in the watershed. Polynomial regression analyses were performed using pollutant sources and impervious coverage data for each subdrainage basin within Hillsborough County. Figure 5.6 illustrates the polynomial models with total ISA in each subdrainage basin as the independent variable and total loadings of nine different non–point source loadings as dependent variables. The exponential polynomial regression models and corresponding r^2 values indicate strong correlations between ISA and most chemical and biological loadings. Most response functions show large increases when imperviousness exceeds 150 km². The best regression response functions for ISA and metal loadings are also exponentials with relatively lower r^2 values compared with regression results from chemical and biological sources except Zn, BOD, TSS, TDP, and $NO_3 + NO_2$ have almost similar correlation patterns and r^2 values. The response functions of TKN, TN, and oil + grease are also very similar with relatively higher r^2 values. The loading of Zn has the highest r^2 value. These relationships suggest that nonpoint chemical, biological, and metal loadings tend to be amplified as urbanization intensifies. The remote sensing-derived ISA density depicts land use patterns at the level of land parcels. Spatial extents of ISA in drainage basins were closely associated with annual non–point source pollutant loadings.

5.5 Satellite-Derived Water Quality Map

5.5.1 Suspended Sediments in the Lake Chicot, Arkansas

Lake Chicot located in Chicot County in southeastern Arkansas is an oxbow lake that was created over 600 years ago by the meandering of the Mississippi River (Schiebe et al. 1992). The lake has a surface area of 17.2 km² and a shoreline of 58.2 km and is the largest natural lake in Arkansas. The length of the lake has been reduced from 32 km to 25.6 km due to natural and human-induced changes. The average depth varied from over 8 m to approximately

3.5 m by sedimentation processes. The constructions of dams and a combination gravity flow pump facility placed in 1985 divert sediment-laden flows into the Mississippi River and regulate lake levels. These activities resulted in a significant reduction in suspended sediment levels in the southern basin of Lake Chicot.

Combining water quality samples collected from four sites at Lake Chicot and corresponding Landsat-MSS data were used to perform analyses for suspended sediment concentration. Two methods have were tested, including (1) linear interpretation of the water quality data collected before and after satellite flyover and (2) through use of the output estimates obtained from running the water quality model. Based on examination of the results of two methods, it was determined that use of the model estimates was more appropriate.

To conduct correlation analysis, each MSS band was plotted as a function of the measured suspended sediment concentration. As expected, reflectance was lowest with lowest suspended sediment concentrations and then increased to an eventual maximum reflectance or asymptotic saturation level. Schiebe et al. (1992) then used the analytical results to develop a remote sensing data based model of the relationship between the spectral and physical characteristics of the surface water. Their analytical model has the form

$$R_i = B_i \left[1 - e^{(c/S_i)} \right]$$

where:
   R_i is the reflectance in wave band i
   c is suspended sediment concentration
   B_i represents the reflectance saturation level at high suspended sediment
      concentrations in wave band i
   S_i is the concentration parameter equal to the concentration when reflectance
      is 63% of saturation in wave band i

This physical-based reflectance model was developed to connect the optical properties of water and water quality parameters. The model containing statistically determined coefficients (B_i and S_i) was successfully applied to estimate suspended sediment concentrations.

5.5.2 Water Properties in Chesapeake Bay

The Chesapeake Bay has highly productive water along the U.S. East Coast region, with about an average flow of 2.3×10^3 m^3·s^{-1} of freshwater from nearby rivers that flow into the Chesapeake Bay, accompanying with dissolved and particulate materials (Schubel and Pritchard 1986). The largest source of nitrogen entering into the Bay comes from agricultural runoff. However, as the population in the watershed increases, many developments expand from city centers to suburban to build up bigger houses on larger lots. Sprawling low-density residential and commercial areas result in additional infrastructure such as roads and shopping centers, increases storm water pollutant loading, and degrades the health water system (Xian et al. 2007). Bio-optical properties of the Chesapeake Bay water are strongly influenced by complex constituents of high phytoplankton concentration, DOM, and total suspended sediment. This study employed the MODIS-aqua-derived ocean color products and a regional total suspended sediment algorithm to evaluate Chl-a and TSS of the Chesapeake Bay (Song and Wang 2012).

Data used for this study include MODIS Level-2 ocean color products, which were generated using the NIR and shortwave infrared (SWIR) combined atmospheric correction algorithm and *in situ* radiometric measurements from 2002 to 2010 from the NASA SeaWiFS Bio-optical Archive and Storage System database for the Chesapeake Bay. Pixels from a 5 × 5 box centered at a location of *in situ* measurements were extracted from MODIS Level-2 data processed with the NIR-SWIR atmospheric correction algorithms and compared to *in situ* measurements for the data matchup analysis. Generally, the MODIS-Aqua NIR and SWIR derived normalized water-leaving radiances [nLw(λ)] data are well correlated to those from *in situ* measurements for most wavelengths.

The satellite derived and *in situ*-measured nLw(λ) were compared at wavelengths of 412, 443, 488, 531, 551, and 667 nm. Generally, the MODIS NIR-SWIR-derived nLw(λ) data correlated well with those from *in situ* measurements for most wavelengths. Specifically, the MODIS-derived nLw(λ) data at 488, 531, and 551 nm are well correlated to the *in situ* nLw(λ) measurements and the matchup comparison shows some noises at wavelengths of 412 and 667 nm. Furthermore, the comparison results suggested that the MODIS NIR R2009 nLw(λ) data are underestimated for most wavelengths. Mean values of the MODIS-Aqua NIR R2009 nLw(λ) at wavelengths of 412, 443, 488, 531, 547 (551), and 667 nm are 0.50, 0.67, 0.91, 1.02, 1.01, and 0.21 mW·cm^{-2}·µm^{-1}·sr^{-1}, respectively, while the corresponding *in situ* mean values are 0.68, 0.93, 1.31, 1.43, 1.45, and 0.40 mW·cm^{-2}·µm^{-1}·sr^{-1}, respectively. In most of other wavelengths, mean values of MODIS-Aqua NIR-SWIR-derived nLw(λ) data are relatively closer to those from the *in situ* measurements, although some slight overestimations for NIR-SWIR-derived nLw(λ) are observed at 412 and 443 nm. The mean ratio of MODIS to *in situ* nLw(λ) measurements show certain improvements in the MODIS NIR-SWIR-derived nLw(λ) data (mean ratios of 0.87–1.29, depending on the wavelength), compared with the MODIS NIR-derived nLw(λ) data (mean ratios of 0.57–0.75). The comparisons between the MODIS-derived Chl-a and the *in situ* Chl-a measurements from three subregions of Chesapeake Bay illustrate MODIS NIR-SWIR-derived nLw(λ) are well correlated to these from *in situ* measurements.

TSS concentration and *in situ* measurements of TSS and K_d(PAR) (water diffuse attenuation coefficient for the downwelling photosynthetically available radiation) with more than ~15,700 data were analyzed using data acquired in the main stem of the Chesapeake Bay from 1984 to 2010. The K_d(PAR) data were converted to the diffuse attenuation coefficient at the wavelength of 490 nm, K_d(490), using a relationship for the Chesapeake Bay (Wang et al. 2009). A regional TSS algorithm was derived for satellite data application by comparing *in situ* TSS data with *in situ* K_d(490) data. The comparison results show that TSS data are strongly correlated to the K_d(490) in the Chesapeake Bay with a correlation coefficient of 0.877. Therefore, TSS can be linearly related to K_d(490) in the Chesapeake Bay as

$$\text{TSS} = 1.7 + 5.263 \times K_d\left(490\right) \text{ mgl}^{-1} \tag{5.10}$$

where:
K_d(490) is in m^{-1}

The linear relationship can be applied to the MODIS-Aqua K_d(490) data estimate TSS concentration for turbid coastal water.

5.5.2.1 Satellite Chl-a Composite Images

Seasonal climatology images between July 2002 and December 2010 from the MODIS-Aqua-derived Chl-a data were produced using the NIR-SWIR method and the standard OC3 Chl-a algorithm (O'Reilly et al. 1998) for the main stem of the Chesapeake Bay. Figure 5.7 illustrates locations of upper, middle, and lower Bays in the Chesapeake Bay region. Climatology Chl-a images in the Chesapeake Bay for spring, summer, fall, and winter suggest that both spatial distributions of the MODIS seasonal climatology Chl-a images and the MODIS nLw(λ) images are very similar, showing that high Chl-a concentrations are in Upper Bay, river branches, and along the coastal lines, while the lower concentrations are in the central part of Middle to Lower Bays and outside of the Chesapeake Bay. Overall, peak Chl-a values are observed in the spring and the lowest Chl-a concentrations are seen in the fall in regions of Upper-Middle Bays and winter in Middle–Lower Bays. Also, Chl-a values are usually overestimated in the Upper and Middle Bay regions but the overall estimate for the regional is still reasonable.

5.5.2.2 Satellite Images and Time Series of MODIS TSS Measurements

Seasonal climatology images between July 2002 and December 2010 from the MODIS-Aqua-derived TSS data using the NIR-SWIR method and a regional TSS algorithm defined in Equation 5.10 were developed for the Chesapeake Bay. The spatial distribution of TSS concentration shows that the highest TSS values appear in Upper Bay, western branches, and along the coasts, while lower values are in the central part of the Middle to Lower Bays (particularly in the western section of the main stem region) in all seasons. The seasonal

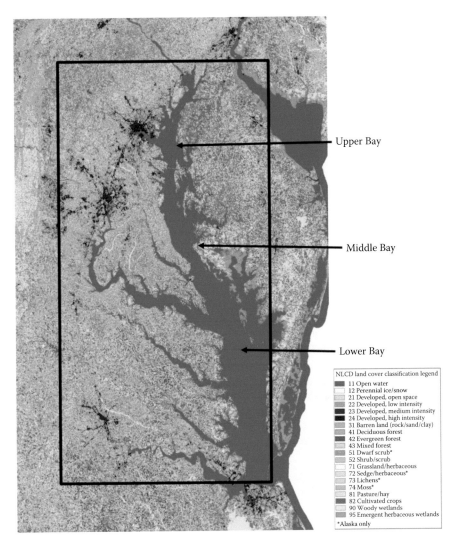

FIGURE 5.7
Land cover around the Chesapeake Bay and adjacent areas for water quality map. Land cover is from U.S. Geological Survey National Land Cover Database (NLCD) 2011.

distribution indicates that the highest TSS values appeared in winter from December to February, especially in Upper Bay. Relatively high TSS values are seen in the eastern area of Middle Bay and various branches. In the spring season, the TSS spatial pattern is similar to that in winter, displaying high TSS values in Upper Bay and various branches and low values in Middle and Lower Bays. Nevertheless, some obviously reduced TSS values are detected in spring, particularly in the central part of Middle to Lower Bays. The lowest TSS values are in summer over all areas. The lowest TSS is geographically located in the western part of Middle Bay and the southeastern part of Lower Bay. The TSS values are slightly increased in the fall season compared to those in summer.

5.5.3 Water Quality Assessment for China's Inland Lake Taihu

Lake Taihu is located in the Yangtze River delta in the eastern coastal region of China. This is one of the world's most heavily populated regions with the highest rate of economic development in recent years. Lake Taihu provides normal water usage for several million residents in nearby Wuxi City (Hu et al. 2010; Wang et al. 2011). Hence, water quality in inland freshwater lakes such as Lake Taihu is vital to human needs and also to the health of regional ecosystems. Lake Taihu, which is the third largest inland freshwater lake in China, has an areal coverage of approximately 2300 km² and a mean water depth of about 2 m. Lake Taihu's water are consistently highly turbid with aquatic vegetation, covering the bottom except in East Taihu Bay and in the part of East Lake regions where water bodies are often clear. Also, Lake Taihu experiences frequent algal blooms most in the spring-summer, resulting in lake water being polluted during these times. For instance, a large-scale blue-green algal (*Microcystis*) bloom occurred in Lake Taihu in the spring of 2007 (Wang and Shi 2008) and caught a wide attention. Algae-polluted water in the lake have impacted the normal life of the several million residents nearby. To effectively monitor and improve water quality for the lake, information about the sources and distributions of pollutant is urgently needed. Figure 5.8 outlines land cover around Lake Taihu where dominated land cover is agricultural and forest lands. The region also has a very high-density population and concentrated built-up lands are observed around the lake.

In the study conducted by Wang et al. (2011), optical and bio-optical properties of Lake Taihu water was characterized using measurements from the MODIS onboard the Aqua satellite platform between 2002 and 2008. Two MODIS NIR bands (748 and 869 nm) were used for atmospheric correction to generate the standard ocean color products with the SWIR-based atmospheric correction algorithm with an iterative approach has been developed specifically for Lake Taihu in order to account for some extremely turbid water conditions and to derive reasonably accurate normalized water-leaving radiance spectra. Specifically, the following procedures have been performed to develop the normalized water-leaving radiance spectra nLw(λ) or the normalized water-leaving reflectance spectra ρwN(λ) over Lake Taihu.

Procedure 1: The SWIR-based atmospheric correction algorithm with use of MODIS-Aqua SWIR bands of 1240 and 2130 nm is processed. Both water-optical property—for example, nLw(λ)—and aerosol property, specifically aerosol Ångström exponents, or aerosol models, are produced at the pixel-by-pixel level. The Ångström exponent $\alpha(\lambda)$ is determined by

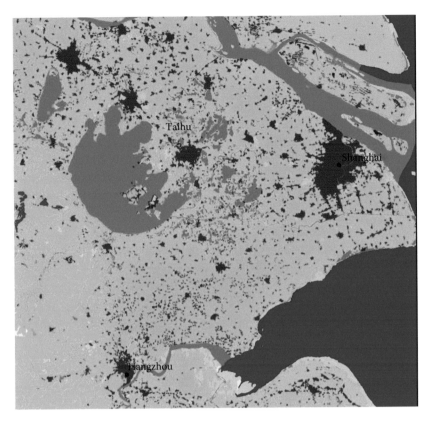

FIGURE 5.8
Land cover map of China's inland Lake Taihu. Colors represent different land cover types: brown is for agricultural land, red is for urban, green is for forest, and light blue is for inland lake.

$$\alpha(\lambda) = \frac{\log_e \left[\tau_a(\lambda)/\tau_a(869) \right]}{\log_e(869/\lambda)} \tag{5.11}$$

where:

$\tau_a(\lambda)$ is the aerosol optical thickness at the wavelength λ

For this specific study, MODIS-Aqua-derived $\alpha(531)$ value is used. It is worth to know that in the data processing system, each aerosol model has a unique $\alpha(531)$ value. This is an essential use of the SWIR-based data processing conducted with MODIS SWIR 1240 and 2130 nm bands for Lake Taihu.

Procedure 2: By using the derived aerosol Ångström exponent values for each MODIS-Aqua scene, a mean value from a box of 20 × 20 pixels (about 4 ×4 km² in the remapped image with about 0.25 km spatial resolution) centered around the central lake region is estimated. In the central lake region, we found that the MODIS SWIR 1240 nm band usually is very dark and can be used to derive representative aerosol properties for Lake Taihu (e.g., aerosol Ångström exponents or aerosol models).

Procedure 3: Using the mean aerosol Ångström exponent calculated from Procedure 2, the SWIR atmospheric correction algorithm using a single band at 2130 nm can be processed for the entire lake to derive the normalize water leaving radiance spectra nLw(λ) data, as well as other water properties such as Chl-a concentration and diffuse attenuation coefficient at the wavelength of 490 nm [$K_d(490)$] (Wang et al. 2009).

By comparing with *in situ* measurements, the SWIR-based iterative atmospheric correction algorithm performed reasonably well for these highly turbid inland lake water bodies. Actually, both the *in situ* and MODIS-Aqua data show significantly high NIR water-leaving radiance values, that is, nLw(λ) values at the wavelength of 859 nm; nLw (859) range from ~0.3 to over ~1 mW·cm^{-2}·μm^{-1}·sr^{-1}.

The MODIS-Aqua-derived products were then used to assess the 2007 algal bloom event. High-intensity Chl-a concentration emerged in the far north shore of the lake in middle April, indicating the blue-green algal bloom contamination. The May 7 observation suggested the blooms of algae peaked in east, north, and west shores of the lake. MODIS-Aqua Chl-a and nLw(443) data also show that the blue-green algal bloom contamination started in the first week of April, peaked around May 7, ended in the beginning of June. North side of Lake Taihu is one of the main water supply sources for nearby Wuxi City and one of the regions where water bodies were considerably contaminated by the blue-green algal bloom.

Normalized water-leaving radiance at a wavelength of 645 nm [nLw(645)] was used to connect to the water near-surface TSS concentration. For most of the lake regions, high nLw(645) values were observed in the winter–spring seasons and low values were revealed in the summer–fall seasons. These nLw(645) variations are mainly influenced by changes of the TSS concentration in the lake.

5.6 Summary

The quality of water of many inland and coastal water bodies are deteriorated due to developmental activities. Water quality variables have optical properties that can be inferred from multispectral satellite imagery. Currently, the three most often used satellite sensors for inland water quality assessment include Landsat, MERIS, and MODIS. These images have different spatial, spectral, and temporal resolutions. They can be processed using either model-based approach or empirical approach to derive water quality information, which is important for monitoring water quality of different water bodies near most urban areas. Model-based solutions have the ability to retrieve simultaneously various optical and biogeophysical parameters from the remote sensing reflectance. Their theoretical basis on sound solutions to the radiative transfer equation makes them potentially a more expedient approach.

Examples were included in this chapter to demonstrate the use of remote sensing data to assess water quality. In the first example, information of urban development extent and intensity developed from remotely sensed data was used to evaluate nine non–point source pollutant loadings in the Tampa Bay watershed. In the second example, an analytical study conducted for analysis of suspended sediment in Lake Chicot based on Landsat MSS data was introduced. A remote sensing data based model of the relationship between the spectral and physical characteristics of the surface water was developed. This is still useful to estimate suspended sediment concentrations.

The MODIS-derived water property in the Chesapeake Bay revealed spatial distributions of TSS and Chl-a in the region. Also, the climatology MODIS-derived water property products also revealed the temporal patterns of high values in spring and winter and low values in summer and fall for nLw(λ) and TSS, and highs in spring and summer and the lows in fall and winter for Chl-a.

The MODIS-Aqua-derived products were used to assess water quality in Taihu Lake, China. The 2007 blue-green algal bloom event was analyzed in detail using MODIS-Aqua measurements, demonstrating that satellite water color data can be a useful tool for monitoring and managing lake water quality and can provide better, quicker, and more valuable water information to support improved decisions for local and national management efforts.

Although these examples show the success of using remote sensing to assess water quality, a certain degree of error will unavoidably be associated with any estimates from remote sensing data. An acceptable level of error should be determined using either sensitivity analyses (e.g., Hu 2009) or *in situ* validation. The size of the error could depend on the magnitude of the signal from the instrument design, the data quality, and the degree to which the algorithm is parameterized to connect *in situ* measurements and remote sensing data (Matthews 2011). With current instrumental constraints and allowing for the complexity of detection, it is important to recognize that certain parameters such as aCDOM can be estimated with less confidence than others. With consideration of the error associated with estimates, it is suggested that, in some instances, the information could be viewed as qualitative rather than explicitly quantitative, although quantitatively accurate estimates are desirable. Thus, a great amount of useful information is readily accessible using the empirical approach by accepting a certain degree of error that remains to be improved. Furthermore, remote sensing will play an important role for global assessments of the quality, quantity, and changes occurring in inland fresh and transitional water bodies within the next few decades.

6

Natural Hazard Assessment
for Urban Environments

6.1 Introduction

Urban growth continues in most metropolitan areas around the world. Associated with urbanization is the landscape change from natural land cover types to anthropogenic impervious surfaces, which has been considered as a key indicator of environmental quality (Schueler 1994). While urban expansions continue in many metropolitan areas, many communities experience fatalities and injuries, property damage, and economic and social disruption resulting from different natural hazards such as landslides, floods, debris flows, and extreme heat events. In the last several decades, for example, population in the United States has migrated toward the coasts, concentrating along the earthquake-prone Pacific coast and the hurricane-prone Atlantic and Gulf coasts, and the value of their possessions has increased substantially (Iwan et al. 1999). Population has concentrated in large cities where many infrastructure are highly vulnerable to damage and distribution, and much of their residents are located in areas at risk to natural hazards and disasters. As consequence came into consideration, there is the need to move toward avoiding creation of additional vulnerabilities and reducing existing ones. The devastating outcomes resulting from the impact of hazards such as earthquakes, flash floods, cyclones, volcano eruptions, and tsunamis may be reduced if appropriate disaster risk reduction measures are put in place.

Natural hazards by themselves do not cause disasters; instead, it is the combination of exposure, vulnerability, and ill preparation of populations that results in disaster (McBean and Ajibade 2009). Vulnerability is a function of physical exposure to hazards, sensitivity to the stresses they impose, capacity to adapt to these stresses, susceptibility, fragility, and the lack of resilience in socioeconomic and physical infrastructure. Concerns about possible effects of natural hazards on coastal urban growth have motivated a number of theoretical, modeling, and empirical studies (Metternicht et al. 2005). As one of the types of natural hazards, landslides resulting from a combination of steep slopes, glacial and postglacial soils, and pronounced wet winter season are common in the Seattle area (Shannon and Wilson 2000). Remote sensing information was usually used as a tool for landslide distribution, qualitative, and frequency analyses (Mantovani et al. 1996). Landslide hazard mapping was benefited mostly from information collected using remotely sensed data. The use of remote sensing within the field of natural hazards and disasters has become increasingly common, due in part to increased awareness of environmental issues such as climate change, but also to the increase in geospatial technologies and the ability to provide up-to-date imagery to the public through the media and the Internet (Joyce et al. 2009). As remote sensing technology improved, near-real-time monitoring and visual images have become available to emergency services and the public in the event of a natural disaster. The techniques needed for monitoring and assessing natural hazards and disasters are required to exploit the available data effectively to ensure the best possible intelligence is reaching emergency services and decision makers in a timely manner. Disaster management usually consists of four phases: reduction (mitigation), readiness (preparedness), response, and recovery (Joyce et al. 2009). Remote sensing plays a unique role in each of these phases, though this chapter focuses primarily on its contribution to the response phase. Currently, the most commonly used satellites for hazard mapping and monitoring include WorldView, QuickBird, IKONOS, RapidEys, EO-1, Terra/Aqua, ALOS, SPOT-4 and -5, Landsat Thematic Mapper (TM), and Enhanced Thematic Mapper Plus (ETM+).

In this chapter, we focus on using remote sensing data to analyze impacts of natural hazards in urban area. Several different types of natural hazards and disasters are introduced in the subsequent sections to determine the commonly used image processing techniques. Section 6.2 illustrates urbanization processes and associated land cover transition in the Gulf of Mexico region. Section 6.3 introduces the use of satellite imagery to evaluate flood risks to estimate the scale and cost of urgent repairs of houses and relevant infrastructure in urban environments. Section 6.4 summarizes an integrated study of using urban land cover data derived from remote sensing information for landslide assessment in two coastal urban areas. Section 6.5 is the summary of this chapter.

6.2 Urban Development and Land Cover Transition in the Gulf of Mexico Region

Urban growth is one of the most important trends in land cover change because of the potential consequences in terms of deterioration of natural vegetation and environmental conditions associated with the long-term replacement of urban land cover. Urban development trends are mainly influenced by population growth and economic conditions. In the United

States, the migration of population has been found to be concentrated along the earthquake-prone Pacific coast and the hurricane-prone Atlantic and Gulf coasts (Iwan et al. 1999). Approximately 3% of the U.S. population lives in areas subject to the 1% annual chance coastal flood hazard (Crowell et al. 2010). These population trends have substantially increased urban land cover in coastal areas and therefore put most urbanized coastal areas under increasing pressure from development and population growth. Meanwhile, these coastal environments are also prone to impacts from a variety of natural hazards, including hurricanes, tsunamis, floods, wildfires, and rise in sea levels. The desire of many residents to live close to the water has made people and infrastructure highly vulnerable to these coastal natural hazards. For instance, in the Gulf of Mexico region, the impact of urban growth on the regional ecosystem conditions is complicated by the regional warming trend, which has been observed since 1970 (http://www.globalchange.gov). Land cover and land cover change information are critical to monitoring the loss of nonurban lands and determining vulnerability to climate change impacts, such as rise in sea levels and storm surges, and to identify the resources required to provide urban services, such as water, energy, and transportation infrastructure (Seto and Shepherd 2009).

To evaluate impacts of urban land cover change and associated land cover transition on the regional ecosystems in coastal areas for potential hazard assessment, a change analysis has been performed in the Gulf of Mexico region (Xian et al. 2012; Xian and Homer 2013). The study area encompasses a portion of Georgia and coastal areas of Texas, Louisiana, Mississippi, Alabama, and Florida in the United States (Figure 6.1). Furthermore, the study area was divided into three regions (eastern, central, and western) to reveal urbanization patterns and trends, as well as the overall and regional landscape effects of the urban land cover change. To examine how urbanization changed land cover condition in coastal areas, each region was further subdivided into three buffer zones of 10, 50, and 100 km from the coastline. The total extent of the 100 km zones in the three regions encompasses approximately 313,932 km^2.

Impervious surface and land cover products from U.S. Geological Survey National Land Cover Database 2001 and 2006, which has been illustrated in Chapter 3, were used to assess the spatial extent and intensity of urban development and associated ecological effects in the study region. The land cover transition associated with urban growth in different buffer zones along the coastlines was also analyzed through conducting Chi-square tests by comparing areal changes of different land covers between 2001 and 2006. The test revealed the significance of different land cover types that were converted into urban land in different zones.

Increases of impervious surface area (ISA) associated with urban developments in the region were directly quantified from the updated ISA product. The increments in ISA were about 107, 88, and 220 km^2 in the eastern, central, and western regions, respectively, resulting in growth rates of 1.1%, 0.9%, and 1.4% from 2001 to 2006 in the three regions. The overall annual growth rate

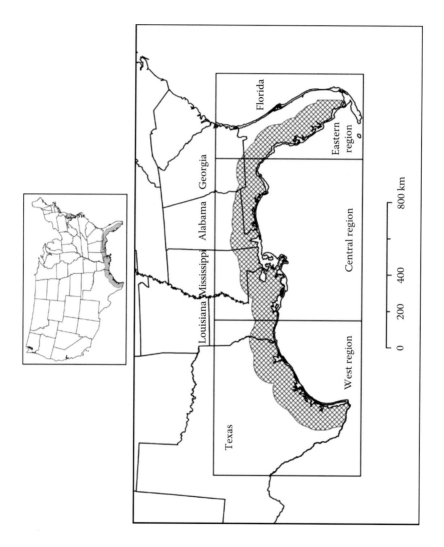

FIGURE 6.1
The Gulf of Mexico region, the 100 km buffer zones, and three subregions of eastern, central, and western regions.

between 2001 and 2006 for the entire study area reached 1.1%, or 5.6% in five years. The change in ISA varies in different zones and regions. In the eastern region, the largest areal increase of ISA emerged in the 10 km zone and the highest growth rate took place in the 10–50 km zone, indicating a large areal increase in urban land cover along the Florida coastal area. The central region has the largest areal increase in the 10–50 km zone and the highest growth rate in the 10 km zone. In the western region, however, the largest areal increase and the highest growth rate of ISA took place in the 50–100 km zone. The region also possesses the largest areal growth and highest growth ratio of ISA in the three regions.

The overall change of urban land cover was totaled from the four urban classes for the study area and both 2001 urban areas and new growths from 2001 to 2006 were illustrated in Figure 6.2. Apparently, most existing urban areas were located in the 100 km zone and most new urban lands emerged around the existing urban boundaries. Associated with urban growths are the transitions of other land cover types to urban land cover. Overall, the leading land cover transitions were from hay/pasture (260 km²) and woody wetland (188 km²) in the 100 km zone over the three regions. The two largest transition rates were 2.2% for barren lands and 0.7% for herbaceous across the three regions. These transition rates varied in different regions and buffer zones. Specifically, the most prevalent land cover transitions from woody wetlands and hay/pasture were observed in the eastern region. The most common transition occurred in hay and pasture and evergreen forest lands in the central and western regions. Overall, 50.8% woody wetland, 48.9% mixed forest, and 53.4% barren-land transitions took place in the 10, 10–50, and 50–100 km zones, respectively, across the three regions.

Significant analysis results obtained from the Chi-square test for all regions and zones regarding the percentage of each urban land cover transition have been presented in Table 6.1. Among these transitions, most transitions in

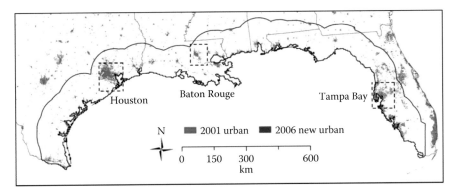

FIGURE 6.2
The 2001 (green) and 2006 new (red) urban areas in the Gulf of Mexico region. Three subset areas in Tampa Bay, Florida; Baton Rouge, Louisiana; and Houston, Texas, are outlined.

TABLE 6.1

Percent of Land Cover Transition between 2001 and 2006 in Different Regions and Zones

Region	Water	Barren Land	Deciduous Forest	Evergreen Forest	Mixed Forest	Shrub	Herbaceous	Hay	Crops	Woody Wetlands	Emergent Wetlands
East	19.9	36**	**1.5***	19.4	**1.7***	12.8*	22.4	25.4	30	43.6***	40.1***
Center	37.1**	20.6	8.8**	33.1	18.2	31	15.5	19.5	29.1	24.8	24.3
West	43.1	43.4	89.7***	47.5	80.1***	56.3	62.2	55.1	40.9	31.6**	35.5*
Buffer Zone											
10 km	39.8**	25	16.1	23.9	**4.1***	20.6	21.1	16.9	13.5*	39.1**	50.8***
10–50 km	30.4	21.6*	41.9	43.2	48.9*	36.5	34.2	33.5	36.6	39.5**	35.6***
50–100 km	29.8	53.4*	42.0	32.9	47.0	42.9	44.7	49.6	50.1	21.4**	13.6***

Note: Significantly higher values than expected are expressed in italics and significantly lower values than expected are expressed in bold.
*p < .05, **p < .01, and ***p < .001.

forest land took place in the western region. More wetland transitions happened in the eastern region than in the other ones. The transitions of deciduous and mixed forests in the eastern region were significant, although the amounts were small. In contrast, only transitions in deciduous forest and woody water were significant in the central region. Among different zones, transitions from emergent wetlands, woody water, and woody wetlands to urban land cover were large and significant in the 10 km zone. Mixed forest transition was the only significant but small land cover change in the zone. In the 10–50 km zone, the large and significant land cover transitions were mixed forest, woody wetlands, and emergent wetlands. In the 50–100 km zone, the barren land transition was large and significant. From land cover transition analysis, it was concluded that woody wetland and hay and pasture were the two most dominant land cover types that were being replaced by urban land in the region. Furthermore, land cover transition imposed substantial pressure to the regional ecosystems, especially for emergent wetlands in the eastern region and in 10 km zone for all regions. The urban developments have significantly impacted the 10 km zone by reducing the mixed forest land and the western region by removing both deciduous and mixed forests. The regional land cover change data provided valuable information for analyzing of regional ecological and hydroclimatic conditions in the region.

6.3 Flood Risks in Urban Environments Using Medium-Resolution Remote Sensing

Assessing disaster risk boils down to quantifying hazard information and exposure, where exposure refers to the exposed assets (referred to also as *elements at risk*) present in hazard-affected areas of the world, which are subject to potential losses. To a large extent, however, successful strategies for risk and disaster assessment depend on the availability of accurate information presented in an appropriate and timely manner. Development in remote sensing techniques has evolved from optical to radar remote sensing, which has provided an all-weather capability compared to the optical sensors for the purpose disaster risk assessment. As different remote sensors have a different ground resolution and varying flood/water detection potential, it is important to be aware of such limitations when processing the respective flood masks when generating flood maps. Usually, using passive remote sensing to map and monitor water inundation can be a challenging process due to the often-coincident cloud cover, and the fact that water is generally not visible under a closed vegetation canopy. The utility of high-temporal-resolution sensors such as advanced very high resolution radiometer (AVHRR) could offer a greater probability of obtaining cloud-free imagery due to its frequent revisit time.

Generally, several main questions need to be answered for the decision-making process regarding the assessment and management of flood risks in urbanized areas. These questions and answers that can be served as a guideline to show the capabilities of remote sensing to contribute to the risk assessment (Taubenböck et al. 2011) include the following:

1. Where are the exposed areas?
2. What and who would be affected?
3. How great will the damage be?
4. Which immediate reaction is necessary to avoid potential risks?
5. How are damage grades classified?

Problems involved in hazard identification, risk assessment, and developing mitigation solutions are spatially related to land development and distribution. Different remote sensing data have been used in the field of flood risk assessment and management, including National Oceanic and Atmospheric Administration (NOAA)-AVHRR images (Islam and Sado 2002), MODIS, Landsat, and other high-resolution sensors (Al-Khudhairy et al. 2005; Taubenböck et. al. 2011).

The first example is from the evaluation of potential exposed area at the megacity of Cairo, Egypt (Taubenböck et al. 2011). The digital elevation model (DEM) data was used to assess the spatial distribution of potentially endangered flood areas and Landsat imagery was used to outline urban footprint. The course of the river is derived from the Landsat ETM+ sensor using an object-oriented approach combining spectral information with shape and context information. Then, the surface height values relative to the river elevation are calculated. The relative height differences between the river and its surroundings in combination with neighborhood information and a region-growing algorithm with surface height were utilized to identify endangered lowland areas. Furthermore, with use of relative high resolution remote sensing and accurate elevation data, they assessed scenario-based flood-prone areas and to obtain a first basic and coarse overview on the hazard perspective. Also, from the vulnerability perspective, areas experiencing continuous urbanization face the risk of flooding considerably. Urban built-up areas were characterized from Landsat data to measure the changes in extent, direction, speed, and pattern of the urban extensions over time. A variety of methods can be used for this purpose. Here, an object-oriented, fuzzy-based methodology was implemented to combine spectral features with shape, neighborhood, and context and texture features. The study shows that the highly dynamic process of spatial urbanization from 1972 to 2008 at the megacity of Cairo, giving insight into spatial expansion of potentially vulnerable areas. Overlapping the scenario-based spatial hazard impact map and urban footprint level helps in the detection of vulnerable areas on urban footprint level and the identification and the quantification of the locations of potentially affected areas and their extents.

The second example is to estimate flooded area using satellite imagery. NOAA-AVHRR images were used to estimate the flood effects of 1988, 1995, and 1998, respectively, in Bangladesh (Islam and Sado 2002). To differentiate between water and nonwater, iterative self-organizing data analysis technique clustering was conducted through unsupervised classification. Three classes of water, nonwater, and cloud were aggregated from the initial several categories. After that, cloud pixels were separated and the three categories were divided into two categories of water and nonwater. By comparing with classification maps, flooded areas were estimated as percentages of the total area after subtracting the normal water area such as rivers, lakes, and ponds. The flood-affected frequency was also estimated by using the images in 1988, 1995, and 1998. The different degrees of the flood-affected frequency are categorized by low, medium, and high hazard areas and are illustrated as follows. The inundated area that did not appear in any of the images in these three times was considered to be a nonhazard area, and the area that appeared in a single image was considered to be a low hazard area. The common inundated area that appeared in two images, and the area that appeared in all three images were considered to be medium hazard and high hazard areas, respectively.

The flood-affected frequency is determined for each pixel as the ratio of the number of NOAA images within the three flood events of 1988, 1995, and 1998, showing inundation to the total number of cloud-free NOAA images available near the peak flood time. The flood-affected frequency for the area flooded for the three events was categorized as nonhazard, low, medium, and high, according to the number of floods.

Flood depth categories are also characterized from remote sensing imagery with use of the maximum likelihood method of supervised classification for the individual pixels. Flood depths were classified as shallow, medium, and deep by selecting training areas of shallow, medium, and deep floods on each image, according to the visual interpretation of differences in color and grayscales for different categories of depth for supervised classification. The average albedo of the pixels with the same flood depth category was estimated and used to examine results of the different categories of flood depth. Furthermore, the ranking for the flood depth was categorized as no flooding, shallow, medium, and deep flooding. After that, three flood depth maps were constructed for the three flood events of 1988, 1995, and 1998.

The last assessment for flood risk is hazard rank (HR) through land cover and geology conditions. A weighted score is developed for each pixels of the land area by

$$\text{Weighted score} = (0.0 \times A) + (1.0 \times B) + (3.0 \times C) + (5.0 \times D)$$

where:

A, B, C, and D represent the occupied area percentage by nonhazardous area, and low, medium, and high flood-affected frequency, respectively, when flood-affected frequency was considered to be a hydraulic factor, for each land cover category

Totally, nine different land cover categories were used to calculate the weighted score. These land cover categories are cultivated land with scattered settlements, rice field, cultivated lowland with scattered settlements, dry fallows, mixed cropped area with scattered settlements, mangrove area, highland with mixed forest, highland with scattered settlements, and saline areas. Similarly, A, B, C, and D represent the occupied area percentage by nonflooded area, shallow, medium, and deep flooding, respectively, for each land cover category, when flood depth was considered to be a hydraulic factor. The acquired area percentages by nonhazardous, low, medium, and high damage areas are only for these nine land cover categories with weighted score and HR. Points for the categories of land cover were estimated on the basis of linear interpolation between 0 and 100, where 0 corresponds to the lowest (0), and 100 corresponds to the highest (230.21) weighted score. The three rankings for flood damage (HR 1–3) were obtained from the allocated point and were used to quantify the flood hazard. These HRs were fixed according to the corresponding value of the points, in which points 0–33, 33–66, and 66–100 represent the HRs 1, 2, and 3, respectively. Additionally, four different categories of flood depth of no flooding, shallow, medium, and deep depth were estimated independently using the same images in three times. HRs were determined for each event using the above-mentioned algorithm different land cover categories by using the flood depth.

It is worth to mention that there has been improvement in the temporal and spatial resolutions of very high-resolution optical imagery for disaster risk assessments. Images from IKONOS (0.82 m resolution in the panchromatic mode and 3.2 m in multispectral bands), QuickBird (0.55 m resolution in the panchromatic mode and 2.44 m in multispectral bands), and WorldView-2 (0.46 m resolution in the panchromatic mode and 1.85 m in multispectral bands) provide the possibility to recognize individual buildings, houses, and other settlements for physical exposure and disaster risk (Ehrlich and Tenerelli 2013) or even structural damage assessments (Al-Khudhairy et al. 2005).

6.4 Landslide Assessment in Coastal Urban Areas

Landslides and floods are common in coastal and mountain areas. Individual floods and landslides can cause serious casualties and damages in many urban areas, especially in many coastal and mountain areas. Landslide events that are usually associated with debris flows are common in coastal and mountainous regions (Sidle and Ochiai 2006). These events are the consequence of a complex interaction between environmental factors and human activities. Intensive, relatively infrequent rainstorms falling onto a previously saturated landscape, the bursting of natural dam formed by landslide debris, glacial moraines or glacier ice, and earthquake shaking or ice melting are main environmental factors that trigger most landslides (Dai et al. 2002; Lorente et al. 2002). Human activities, including construction, excavations, mining, and deforestation, also contribute to the triggering of harmful landslides (Guzzetti et al. 2005). The potentially destructive effect of artificial structures and poor planning in design and management of man-made infrastructure can result in numerous casualties. Many elements of these complex infrastructure are highly vulnerable to breakdowns, which can be triggered by relatively minor events. Vulnerability also depends on many other factors, including magnitude of hazard; timing, persistence, and reversibility of impact; and estimation and perception of risk. It is vital for many coastal cities to have updated and accurate urban land cover and infrastructural information to develop strategy for reducing losses caused by natural hazards. Accurate and relevant information can be used to substantially reduce potential losses in many threatening situations.

According to a study conducted in Italy, in the 686-year period from 1317 to 2002, the historical catalog of floods with human consequences lists 1019 events with casualties (Guzzetti et al. 2005). Flood fatalities have counted 36,020 in 837 fatal events, equivalent to a frequency of 1.22 fatal events every year. In the period between 1900 and 2002, the flood catalog records 893 events, in which 755 were related to fatalities, corresponding to a frequency of 7.33 fatal events every year. The average number of fatalities caused by an inundation is 42.9 with a frequency of 3.8 in the period between 1900 and 2002.

Concerns about possible effects of natural hazards on coastal urban growth have motivated landslide study in this section. It is possible to use both radar and optical remote sensing data for landslide detection. However, optical data provides better results, most likely due to spatial resolution and sensor look angle. It can suffer from misclassification with other areas of bare ground. Multitemporal analysis is preferable and spectral enhancement is often required. In this section, two examples of landslide assessment using remote sensing are introduced.

6.4.1 Detection of Landslides Using Remote Sensing in Hong Kong

Hong Kong is one of the metropolitan areas that has the highest urban population densities in the world. Nearly 40% of the land area of the Hong Kong Special Administrative Region is designated as country parks (Nichol and Wong 2005a). A rainfall event affecting 20%–50% of Hong Kong has the potential to trigger a high density (>10 km^{-2}) of natural terrain landslides in susceptible areas. With consideration of both social and economic effects, landslide hazard assessment is an important factor in any development project. Nichol and Wong (2005a, 2005b) have used both high-resolution IKONOS and medium-resolution SPOT XS data with a landslide database created by the Hong Kong government to detect and interpret landslides in Hong Kong.

The Hong Kong landslide study was performed in a 36 km² landslide-prone area on Lantau Island, Hong Kong. The terrain is mountainous with steep slopes, attaining heights of over 800 m only 3 km inland from the coast (Figure 6.3). The main vegetation types include lowland forest in the valleys with shrub and grassland at the higher levels and on ridges and summits.

To cover the area, two SPOT images, on December 21, 1991, and February 5, 1995, were acquired. The images were taken in pre- and postdate of a serious rainstorm that induced landslide event in November 1993, wherein 551 landslides were observed.

Change detection is based on single change pixels and can operate at pixel or even at subpixel level, given the high contrast between the object (landslide) and its background. Planning data in a GIS are available to verify change pixels that are not actually landslides are due to human development such as roads and buildings in the study area.

The maximum likelihood classifier approach was found to be the most accurate and objective and was used for change detection. For the maximum likelihood classifier, all significant land cover types in the image were characterized as different classes, including shadow, which was a significant component of the images due to the steepness of terrain combined with the low sun angle in winter. It was included as a class to reduce the classification

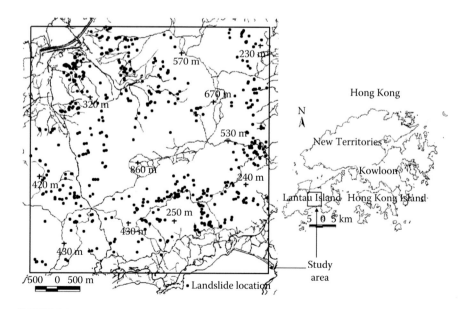

FIGURE 6.3
Location of the study area in Lantau Island, Hong Kong, and the distributions of landslides reported between 1991 and 1995. (From Nichol, J., and Wong, M.S., *Inter. J. Remote Sens.*, 26, 1913–1926, 2005b, Figure 1. With permission from Taylor & Francis Group.)

confusion between dark shadowed areas and forest and water, which had similar spectral characteristics. The overall classification accuracy that was assessed by comparison of 100 random sample points with the orthorectified air photographs showed the overall class accuracies of 85% for the 1991 classified image and 87% for the 1995 classified image. The classified images for each date were combined together to produce two change images to represent pixels that changed from grassland to soil, and those that changed from woodland to soil over the time period. This was done for each pixel using the following conditional filters:

IF input91 = grassland AND IF input95 = soil THEN output = grassland
   _to_soil

IF input91 = woodland AND IF input95 = soil THEN output = woodland
   to soil

Then, the Natural Terrain Landslide Inventory data were overlaid onto each change image and the landslides whose crowns and/or trails were seen to be overlapping by more than 60% with change pixels were flagged as successfully detected. Among the 551 landslides occurring in the study area between the two image dates, 75 were in grassland and 59 in bare soil areas on the upper slopes. The remaining 417 occurred in woodland on midslopes, which tend to be the steepest. The image change detection indicated that 67% of landslides in grassland areas and 71% in woodland areas were detected. Figure 6.4 shows the appearance of landslides with crowns and trails of different widths on the same 1:10,000 scale air photographs. These

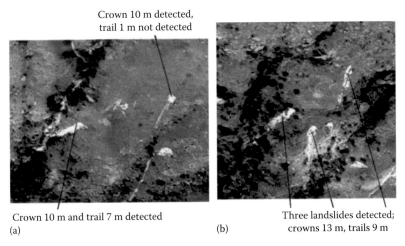

Crown 10 m detected,
trail 1 m not detected

Crown 10 m and trail 7 m detected

(a)

(b)

Three landslides detected;
crowns 13 m, trails 9 m

FIGURE 6.4
Landslide scars and their width extracted from 1:10,000 scale air photographs (a) and SPOT change images (b). (From Nichol, J., and Wong, M.S., *Inter. J. Remote Sens.*, 26, 1913–1926, 2005b, Figure 2. With permission from Taylor & Francis Group.)

were matched visually with the SPOT change pixels and their detectability is stated on the figure. Generally, crowns and trails as small as 7–10 m width were detected on the SPOT change images. In Figure 6.4a and b, five landslides having crowns and trails between 7 and 13 m wide are displayed and all of which are detected on the SPOT change images. However, a 1 m wide trail (Figure 6.4b) was not detected. Thus, the threshold of detectability of landslides on the SPOT images appears to be at subpixel level.

6.4.2 Assessment of Landslide Environment in the Seattle Area

We focus on using Landsat satellite data to estimate multiyear impervious surfaces conditions in Seattle, Washington, and connect land cover change with landslide events in the region to evaluate environmental changes in areas experiencing landslides.

The Seattle–Tacoma region lies in the northwest corner of the continental United States, on Puget Sound in western Washington. This area extends about 140 km north to south and 60 km east to west. It encompasses approximately 6700 km² of land in Island, King, Kitsap, and Pierce counties (Figure 6.5). Major cities include Seattle, Tacoma, Bremerton, and Marysville. More than three million people reside in the region.

The Seattle–Tacoma area's temperate climate and growing economy have led to the cities being ranked as some of most livable areas in the United States. The regional climate and economic growth attracted more people to move to the area. Urban development has been tremendous in the last 50 years. The total population of these cities has almost doubled since 1965. In King County, between 1970 and 2000, for example, the population increased 44%, from 1.2 to 1.7 million, while the number of households increased by 72%, from 400,000 to 680,000 (KCPSB 2010). The City of Seattle is located in the Puget Lowland basin between the Olympic Mountains to the west and the active volcanic arc of the Cascade Range to the east. The basin is characterized by the rolling northeast-oriented ridges left by the last glaciations about 15,000 years ago (Booth 1987). Relief in the area is modest with elevations extending to about 500 m.

Landslides triggered by precipitation and earthquakes were a recurring problem on many hillslopes in the Seattle area. Most landslides commonly occurred at and near the contact between relatively permeable, advance outwash deposits of sand and gravel of the Vashon stade and underlying, relatively impermeable lacustrine beds of fine-grained, clayey silt or clay-rich, pre-Fraser sediments (Coe et al. 2004). Landslide records during the period from 1890 to March 2003 organized by the City of Seattle showed that a total of 1433 landslides occurred during that period.

All recorded landslides were identified as having natural, human, or unknown triggering mechanisms (Laprade et al. 2000). Records from 1890 to 2000 indicated that 87% of the landslides had natural triggers (precipitation or earthquake), 7.5% had human triggers, and 5.5% had unknown triggers (Laprade et al. 2000). Historical records suggested that most landslides in the

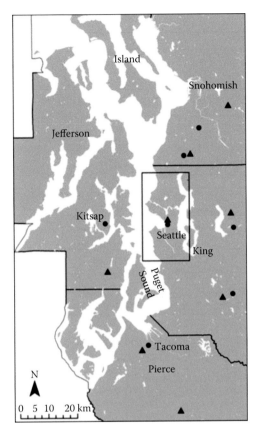

FIGURE 6.5
Seattle–Tacoma area in western Washington State. The triangle marks and solid circles represent locations of selected orthoimages for training and validation datasets, respectively. The area outlined by a rectangle is the City of Seattle, Washington.

area were trigged by heavy rain and snow falls. Precipitation is generated almost entirely from wintertime; cyclonic storms and convective storms that usually produce intense rainfall are rare in the region (Church 1974; McGuirk 1982). Approximately 72% of precipitation comes in the form of drizzle with rainfall of intensity less than 1 mm/h and less than 5% of precipitation comes in the form of snow or other frozen precipitation (Church 1974). When snow falls, however, the melting snow mixed with rainfall can trigger widespread landslides. Landslides triggered by precipitation in Seattle occurred almost entirely in the winter season. As population growth continues in the region, urban expansion into hillside areas is placing an increasing number of people and property at risk from landslides. It is vital for coastal cities such as Seattle to have updated and accurate urban land cover and land use information to develop strategies for reducing losses caused by natural hazards such as landslides.

6.4.2.1 Remote Sensing Data

To characterize regional urban land cover conditions, the approach introduced in Chapters 2 and 3 has been implemented. Remote sensing data include both high-resolution digital orthophoto quarter quadrangles (DOQQs) and Landsat TM images. Eight 0.3 m orthoimages acquired from eight different locations in the Seattle metropolitan area were selected. Each Seattle ortho-image covers approximately 1.6×1.6 km². Four Landsat scenes acquired in 1986, 1990, and 1994, and ETM+ images in 2000 and 2002 were used to quantify urban land cover changes.

To reduce atmospheric effects, the histogram matching method was used here to normalize images of the same scene taken at different times. This was accomplished by matching the histograms of each spectral band of images in different years to the corresponding histogram of the 2002 image. The spectral variations for the same type of land use were adjusted to have similar response patterns for observations in different dates. One side effect associated with this process is that it can mask real changes between the images because the histogram matching technique is a simple numerical fit (Schott 1997). However, this masking effect was minimized when the differences between the scenes only represented a small number of pixels and the transforms were based on whole-image statistics.

6.4.2.2 Impervious Surface Estimate

The orthoimages were characterized as urban and nonurban using supervised classification. After the orthoimages were classified, they were processed to calculate percent imperviousness, which was used as training data to build regression tree models by means of regression tree algorithms.

Landsat reflectance, normalized difference vegetation index (NDVI) derived from Landsat reflectance, Landsat thermal bands, and slope information were all input as independent variables to build regression tree models. However, because some Landsat images were acquired prior to orthoimages, land cover condition could not be the same in Landsat imagery as captured in orthoimages. To minimize the impacts caused by landscape changes, orthoimages were carefully inspected through comparing with Landsat images to remove the training data that did not match landscape conditions in a specific year. Then, different numbers of training datasets estimated from the 2003 orthoimage were selected to build the training dataset for Landsat scale impervious surface estimate for different years. For example, if the area in the orthoimage had apparent changes when compared to the Landsat image of a specific year, the training impervious surface calculated from the orthoimage was excluded from the training data poll. Therefore, the actual numbers of imperviousness samples calculated from orthoimages and used for the training dataset were six for 1986 and 1990, seven for 1994, and eight for 2000 and 2002 impervious surface estimations. Subsequently, percent impervious surfaces in the Seattle area were estimated using training datasets, Landsat imagery, and ancillary data in corresponding times in 1986, 1990, 1994, 2000, and 2002.

Figure 6.6 represents Landsat images and the spatial distributions of ISA in 1986, 1990, 1994, 2000, and 2002 in the region. The Puget Sound metro region is the most developed region in the Seattle area with total areas of imperviousness from approximately 1285 km^2 in 1986 to 2007 km^2 in 2002. Most new developments after 1986 were observed on the eastern side of Seattle and the southern part of Tacoma in King and Pierce Counties. Major increases of ISA associated with urban land use expansions were seen after 1994 in east and southeast of Seattle, and south and southeast of Tacoma. The east coast of Puget Sound was also developed to contain many medium-density built-up lands. The most dominant changes in impervious surface emerged between 20% and 59%, which belonged to low- to medium-density urban land cover categories. The land use pattern and change trend were highly related to the regional geographic condition and development policy. For example, growth management efforts were implemented by King County with a comprehensive plan that brought serious growth management efforts in the 1985 and 1994 to manage new growth while meeting economic needs and providing

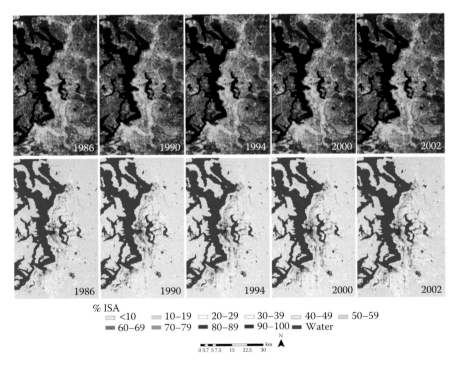

FIGURE 6.6
Landsat images (upper panels) and impervious surfaces (lower panels) in the Seattle region between 1986 and 2002.

affordable housing (Robinson et al. 2005). The main goal of the plan was to encourage development in urban areas and discourage inappropriate low-density development. Residential developments in existing urban areas were zoned for higher residential densities, usually 1–12 dwelling units per acre, while rural areas were zoned for lower residential densities, generally 1 dwelling unit per 2.5–10 acres (KCORPP 2010). As a result, urban growth planning in King County had led to a dual development: most new growth desired over a 20-year planning period was to be within urban growth boundaries, while low-density residential zoning and long-time resource production lands were planned to reduce the potential of new developments outside the urban growth boundaries. Therefore, a wide spread of medium- to low-density built-up lands were observed in the rural residential area. The modeled ISA characterized features of urban built-up land in both spatial extents and temporal variations in the region. Furthermore, the remote sensing derived ISA data also revealed the complexity of urban development patterns in the Seattle area.

6.4.2.3 Evolution of Landslides in the Seattle Urban Area

Landslides induced by precipitation occurred more frequently than other types of landslides (Coe et al. 2004) in the Seattle area. Figure 6.7 displays the spatial locations of recorded landslides and topography conditions in this area. Many of the landslides, occurring during or immediately following heavy rainfall or snowmelt in the area, were shallow failures of colluvium mantling the slops along Puget Sound (Godt et al. 2006). In addition, the desire of homeowners for building houses with views of Puget Sound and the location of transportation infrastructure have placed people and their properties in hazardous areas on and near steep coastal bluffs. Many elements of these constructions were highly vulnerable to breakdowns, which can be trigged by relatively minor events. In the winter of 1996 to 1997, landslides trigged by precipitation caused fatalities and widespread damages within the City of Seattle (Grestel et al. 1997). The damage to city facilities alone exceeded $34 million (Paegeler 1998).

Mounting losses in human casualties and property damage motivated many efforts in risk assessment and prediction, such as probabilistic approach based on the historical record of landslides and empirical approach using 25 years of hourly rainfall data and a record of landslide occurrence. In addition, information regarding geological conditions such as slope and soil condition was also important in localizing the occurrence of landslides in Seattle (Coe et al. 2004). The analysis of landslide locations and associated slopes derived from a 10 m DEM indicates that approximately 90% of historical landslides took place on slopes greater than 8 degrees (a 14% slope) (Coe et al. 2004).

To look at temporal variations of landslide occurrence in Seattle, historical landslide records from the City of Seattle were analyzed. The reliable monthly time-of-occurrence information was available starting in November 1909. For this reason, and because only two landslides were recorded between 1890 and 1909, only records after 1909 were used. The number of landslide

FIGURE 6.7
Landslide occurrence locations (black dots) and shaded-relief map of around the City of Seattle. The landslide database contains all occurrences from 1890 to 2003 in the area. Major transportations are shown by solid white lines.

occurrences in each year were displayed as a histogram in Figure 6.8. Two peaks of landslide events that were trigged by two major winter storms in 1986 and 1997 are observed from the graphic. The histogram, however, does not exhibit the change trend because the annual occurrence of landslide might be related to other natural disturbances. To remove the annual fluctuation, records in every year were regrouped to every decade (10 years) from 1900 to 2000 in order to reveal the long-term variation pattern. The dash line in the figure represents a per-decade change trend. An obvious cyclic variation in the number of landslides has transpired from 1910 to 1980, with the number of events increasing in recent decades. A regression analysis was also performed to find the best fit that describes the decade landslide curve. The variation pattern in the last two decades follows closely to a linear increase trend. A regression analysis found the best-fit pattern associated

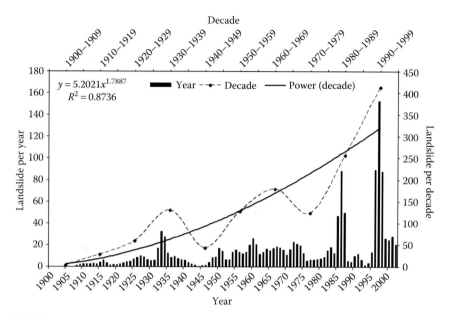

FIGURE 6.8
Multi-year landslide records between 1900 and 2003. The records are grouped in every decade from 1900 to 2000 in order to show long-term trends. A regression line is plotted according to the decade change.

with the decade change trend. The best-fit line displayed in Figure 6.8 is restricted by the following equation:

$$y = 5.2021x^{1.7887}$$

where:

y is the landslide occurrence in every decade
x is time in decade

The regression line has an r-squared value of 0.874. The long time landslide occurrence in the area has an exponential increase trend.

For the area displayed in Figure 6.7, which includes the City of Seattle, the growth of built-up land is the dominant land cover change. Consequently, the change of impervious surface estimated from Landsat imagery suggests that the total urban land cover increases from about 1285 km² in 1986 to about 2007 km² in 2002. The largest increase occurred in low-density urban land use (ISA ~ 10%–40%). By 2002, this land use category accounted for approximately 50% of the total urban land and almost 15% of total land in the region. Medium-density built-up land (ISA ~ 41%–60%) was the second largest category of urban land use, totaling about 424 km² or 6.3% in 1986, increasing to about 635 km² or 9.4% in 2002. High-density built-up land (ISA ≥ 61%) was the smallest urban land use category, encompassing about 252 km² or 3.8% in 1986 and expanding to about 378 km² or 5.6% of total land in 2002.

To investigate the relationship between land use and land cover (LULC) and areas experiencing landslides, quantitative urban land use information was used to measure ISA for those pixels that are associated with landslide events. This analysis was performed on one-time ISA mapping data with three time-frames of landslide data—one year prior to ISA mapping, the year of ISA mapping, and year post-ISA mapping—in order to list the magnitudes of ISA for these pixels. An average percent ISA was calculated by totaling the values for three timeframes above. The three-year moving average is calculated to deter-mine mean ISA in the area that landslides occurred by following formula:

$$\bar{y}(t) = \frac{\sum_{i=1}^{l} ISA(t-1)_i + \sum_{j=1}^{m} ISA(t)_j + \sum_{k=1}^{n} ISA(t+1)_k}{l+m+n}$$

where:

$\bar{y}(t)$ is the average ISA for areas that had landslide impact in year t

$ISA(t)_i$ is the ISA in ith pixel that had landslides in year t

l, m, and n are number of pixels that had landslides in year $t-1$, t, and $t+1$, respectively

Therefore, years $t-1$, t, and $t+1$ represent one year prior to, the year, and the year after ISA was estimated. The mean ISA value represents urban land cover conditions of areas that were impacted by landslides in a specific time by assuming that the built-up land did not have substantial changes in the three-year period and one time ISA status is appropriate to characterize the land cover condition for the period. Thus, these areas can be characterized as hazard impact zones. Figure 6.9 displays the average ISA in 1986, 1990, 1994, 2000, and 2002 for pixels that experienced landslide events. The standard deviation of average ISA in each year and a linear regression line are also embraced in the figure. Magnitudes of the average ISA for landslide impact zones were 8.8% in 1986, 12.4% in 1990, 14.9% in 1994, 18.8% in 2000, and 25.9% in 2002. The change in the relationship of the average ISA line to the standard deviation line indicates that from 1986 to 2002, an increasing proportion of the Seattle urban area has experienced landslides because of increases in both the number of occurrences and the quantity of built-up land. Examination of the Figure 6.9 shows that the highest standard deviation value of 20.7 appears in 1990, which is indicative of a wide range in the number of landslides during that decade. The smallest value of 13.7 emerged in 1986. In 2002, the magnitude was 17.8. There is no appar-ent trend in the standard deviation for the area. Furthermore, means of NDVI in these landslide hazard areas were also calculated using Landsat reflectance bands. The average fractional vegetation cover varies from 42% to 49% in the period between 1986 and 2002, indicating that most landslide events occurred in areas having at least 40% of fractional vegetation cover.

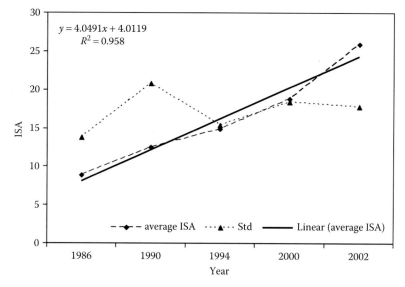

FIGURE 6.9

The average ISA of pixels associated with landslides (long dash line), standard deviation of the average ISA (short dash line), and linear regression line (solid line) for the average ISA, showing an increase trend of imperviousness in areas that experienced landslides.

The assessment of the spatial distribution and temporal variation of landslide suggests that hazard zones can be characterized by both geographic variables, such as slope, soil saturation, and underlying geology, and urban land cover characteristics. The quantitative delineation of urban land cover and its spatial/temporal changes indicate that landslide-prone terrains in Seattle are characterized by increases in impervious surface area and are composed of at least 40% vegetation cover.

As the urban growth continues, the pressure to build in vulnerable areas has also increased. The increase of landslide occurrences from many open spaces to residential areas, or low-density urban built-up lands, indicates more urban developments being built up in hazard zones. Two direct impacts can be related to such developments in these areas. One is the increase of vulnerability for both property damage and public safety. Another impact is the potential increase of landslide due to residential developments, for example, the increase of the quantity of water introduced into the hillslope soil mantle can be greatly increased by irrigation, swimming pools, small artificial ponds, and septic drainage fields (Sidle and Ochiai 2006). The increase of surface runoff due to increasing impervious surfaces can also intensify failures in filling materials and other marginally stable slopes in urban environments. The quantitative information of impervious surface and its changes provided valuable information for landslide occurrence assessments for both current status and long-time evolutions.

6.5 Summary

The use of remote sensing for mapping and monitoring natural hazards in urban areas has diversified in recent years due to an increase in data availability and technological advances in their understanding. Remote sensing has proven useful for a range of applications for natural hazard monitoring and assessment, including the detection of flooding, landslides, and other damages such as earthquakes. In this chapter, applications of remote sensing for natural hazard assessment are introduced. Examples illustrated here suggest that remote sensing information can be used to compare the risk posed by floods, landslides, and other events. Section 6.2 shows how satellite-derived land cover data can be used to assess regional land cover transitions associated with urban land cover change. Such analysis can provide land cover change information in a large geographic extent for natural hazard impact assessments for local, regional, and even national scales. Flood hazard assessments presented in Section 6.3, for example, show that the interactive effect of flood-affected frequency and flood depth were estimated from NOAA-AVHRR images. The floods hazard maps based on each pixel and each administrative district represent the magnitude of flood damage for these areas. This type of map provides capabilities for the responsible authorities to better comprehend the inundation characteristics of the floodplains. In Section 6.4, remote sensing information was demonstrated as a tool for landslide distribution, qualitative, and frequency analyses. To evaluate the landslide in the Hong Kong area, SPOT images were implemented to detect locations and spatial extents of these landslides. The Seattle example revealed that landslide occurrences and their spatial distributions were highly correlated with urban land cover condition in the Seattle area. By using spatial distribution and temporal change of impervious surface derived from remote sensing data, the impact of two-decade landslide occurrences on urban land cover was characterized.

As the importance of good spatial data is becoming increasingly available, remote sensing in the field of hazard assessment and disaster management is likely to grow in the future. New Earth observation satellites are continually being launched, recognizing the prospective market in disaster management. SAR and TerraSAR-X data and techniques, for example, are of considerable value for mapping flooding extent in urban areas. Recent progresses in using these data to flood mapping in urban areas encourage the use of SAR and TerraSAR-X data for natural hazards monitoring, especially the latter one's high resolution in stripmap/spotlight modes (Mason et al. 2010; Giustarini et al. 2013). The flexibility provided by multi-sensor, multiplatform, and quick turnaround interpretation technology of data interpretation are likely to provide the most comprehensive information for the disaster management.

7

Air Quality in Urban Areas—Local and Regional Aspects

7.1 Overview

Urban air pollution is one of the top 15 causes of death and disease globally, as high as in top 10 for high-income countries, responsible for an estimated one million deaths annually (Bechle et al. 2013). Urban areas have unique pollutant properties. Common pollutants in surface air associated with human activities in urban areas include aerosols, ozone (O_3), nitrogen dioxide (NO_2), carbon monoxide (CO), and sulfur dioxide (SO_2). Air pollution generated in many urban regions has become one of the most important environmental problems in the last few decades because of its hazardous effects on human health and its potential impact on local, regional, and global climate.

Atmospheric aerosols or particulate matter (PM) are liquid and solid particles suspended in the atmosphere from natural or human-made sources. These particulates can be a result of anthropogenic processes such as fossil fuel combustion or biomass burning, or can come from natural sources such as sea salt and dust, although some dust emissions can be deemed anthropogenic if they originate from such processes as mining or farming (Coutant et al. 2003). The emission rate of anthropogenic aerosols as well as of aerosols generated by land surface modification, has been increased considerably in many urban areas since the beginning of the industrial era. Aerosols have extensive impacts on our climate and our environment (Kaufman et al. 1990, 2002), particularly, aerosols in the atmospheric boundary layer can bring adverse effects on public health (World Health Organization 2000). PM is a type of air pollution that includes fine dust, dirt, soot, and smoke. It can also appear as droplets. PM is generally grouped into two sizes: PM_{10}, which stands for PM measuring 10 µm or less and thinner than the width of a human hair, and $PM_{2.5}$ that identifies particulates smaller than 2.5 µm. $PM_{2.5}$ emissions are mainly due to wind-blown dust from urban construction, unpaved roads, mining operations, and combustion particles from automobile emissions. These particles can stay airborne several days and make up as much as half the haze in many metropolitan areas. In high enough concentrations, particulates can aggravate existing respiratory problems or even cause severe problems for young and elderly people. Regular exposure to $PM_{2.5}$ has been linked to an increased risk of cardiovascular or respiratory disease, along with lung cancer. $PM_{2.5}$ can also cause a myriad of short-term health issues such as respiratory infections, irritation of the eyes and nose, and headaches (Pope et al. 1995; World Health Organization 2000; Puett et al. 2009). In addition to health concerns, suspended particulates also reduce visibility and contain harmful compounds, which can cause environmental damage via dry or wet deposition. The type of PM and its concentration usually have a large variability in and around highly developed urban areas due to the land use practices that influence the amount of PM by replacing natural land cover with man-made sources of pollution. The conversion of forest, grasslands, and farmland to residential housing, industrial complexes, and large commercial centers often lead to an increase in emissions from industrial activities and automobile emissions.

Ozone is a form of oxygen with three atoms and produced in troposphere by photochemical oxidation of volatile organic compounds (VOCs) and CO with nitrogen oxide radicals ($NO_x = NO + NO_2$). At ground level, ozone is an air pollutant that causes health and environmental problems (Boubel et al. 1994). Ground-level ozone, which is a key ingredient of urban smog, forms during daytime from a complex chemical reaction. Sunlight, heat, NO_x, and VOCs make up the key ingredients of ground-level ozone. Both nitrogen oxides and VOCs are produced when fossil fuels are burned in motor vehicle engines, power plants, and industrial boilers. There are many other sources of VOCs, including gasoline vapors, dry-cleaning products, and chemical solvents. Seasonal weather conditions and traffic congestion during the hottest months of the year are significant factors in the formation of ground-level ozone. Heavy anthropogenic pollutions, which are usually associated with urban growth and industrial activities, can increase ozone concentrations to detrimental levels (Cartalis and Varotsos 1994; Zhang et al. 2004; Xian 2007). People with respiratory illnesses such as asthma, bronchitis, and emphysema are most vulnerable, but when ground-level ozone levels are high, even healthy people who are active outdoors can be affected. Symptoms may include coughing, wheezing, itchy eyes, nasal congestion, and reduced resistance to colds and other infections.

Ground-level air quality has strong local and regional variability because the formation of O_3 and aerosols is a complicated course and depends on the sources of their precursors. Traditionally, observation information collected from surface observation sites are usually used to monitor and analyze local and regional air quality. However, air quality managements in many urban areas are impeded by uncertainty in traditional bottom-up emission inventories. Surface air quality in urban areas has a highly heterogeneous feature in its spatial distribution due to rapid vertical mixing during the day. The long-range transport in the atmosphere of pollutants from rural to urban or from urban to rural area is also an issue of growing concern (Keating and Zhuber 2007; Zhang et al. 2008). These concerns create the need for further development of remote sensing techniques for large-scale air quality assessment and monitoring. Advances in the utilization of satellite data in monitoring aerosol characteristics make it possible to use satellite image as a tool for remote sensing of the atmospheric aerosol and trace gases.

Satellite remote sensing of the lower atmosphere is generally classified into three categories. The majority of instruments employ passive techniques by observing either solar backscatter (<4 μm) or thermal infrared (TIR) emission (4–50 μm). Another type of observation is from active instrument that transmits energy downward and measure the backscatter. In this book, the first two types of instruments are introduced.

Satellite retrievals of the atmospheric aerosol have focused on deriving the aerosol optical depth (AOD) or thickness from satellite visible images. The derivation of AOD is based on the aerosol effect on the upward radiance. The difference between the upward radiance on a hazy day, which has a large optical thickness, and the radiance on a clear day, which has a small optical thickness, can be used to derive the aerosol optical thickness and further to estimate the mass loading and transport. Chemical species in the atmosphere are detected by the absorption, or attenuation, of radiation of specific wavelengths along the path that the radiation travels through the atmosphere.

Satellite remote sensing of trace gases and aerosols for air quality applications can be traced back to the early 1970s when the Geostationary Operational Environmental Satellite (GOES) satellite image was used (Martin 2008). These studies include land use information from the Landsat satellite complemented with ground-based monitors to determine population exposure to air pollution (Todd et al. 1979), GOES observations conducted the first retrieval of AOD over land and applied it to examine a haze event over the eastern United States (Fraser et al. 1984), and O_3 columns were retrieved from the total ozone mapping spectrometer (TOMS) satellite instrument to examine a surface O_3 episode over the eastern United States (Fishman et al. 1987).

Satellite remote sensing of pollutants including O_3 and aerosol precursors provides more broad information related to the spatial distributions on emission inventories through a top-down method built on inverse modeling of observations. In the last two decades, major advances have been achieved in the detection of atmospheric pollution from space. The generation of satellite instruments launched since 1995 has provided capability of observing a wide range of chemical species at increasingly high spatial and temporal resolutions (Streets et al. 2013). In addition, the transformation of raw satellite retrievals to user-friendly, archived products has progressed considerably, so that the application of satellite observations to a wide range of atmospheric problems is feasible. In this chapter, Section 7.2 gives an overview of the major satellite instruments being applied to remote sensing of surface air quality. Section 7.3 introduces medium-resolution remote sensing for air quality assessment. Section 7.4 discusses retrieved species and their application to surface air quality. Section 7.5 presents several applications of surface air quality assessments in some urban areas. Section 7.6 examined urban land impact on air quality. Section 7.7 is the summary of the chapter.

7.2 Satellite Retrievals

Satellite retrievals of air quality can provide information for basic research and operational needs to support air quality management and public health advisories. The majority of instruments of satellite retrievals of the lower atmosphere employ passive techniques, observing either solar backscatter (<4 μm) or TIR emission (4–50 μm). Instruments are characterized into two types: those that observe solar backscatter radiation in the ultraviolet to the visible and those that observe TIR emission.

However, satellite instruments do not directly measure atmospheric composition. Species in the atmosphere are detected by the absorption, or attenuation, of radiation of specific wavelengths along the path that the radiation travels through the atmosphere. The inferred concentration of the chemical species, known as *retrieval*, is determined by a complex set of spectral fitting and radiative transfer calculations (Kaufman et al. 1997; Martin 2008; Streets et al. 2013). Retrieval is usually performed by calculating the atmospheric composition that best reproduces the observed radiation. Such retrievals often require external information on geophysical fields as described in the subsequent sections. The development of a variety of algorithms to extract physical parameters by accounting for atmospheric radiative transfer has been integral to the success of modern remote sensing.

Quantification of emissions of ozone and aerosol precursors for a large spatial extend can be accomplished by satellite remote sensing images due to their large spatial coverage and reliable repeated measurements. The recent advances in tropospheric remote sensing from low Earth orbit instruments such as measurements of pollution in the troposphere (MOPITT), the global ozone monitoring experiment (GOME), moderate-resolution imaging spectroradiometer (MODIS), multiangle imaging spectroradiometer (MISR), Scanning Imaging Absorption spectroMeter for Atmospheric CHartogaphY (SCIAMACHY), ozone monitoring instrument (OMI), and tropospheric emission spectrometer (TES) have demonstrated the value of using satellites for both scientific studies and environmental applications (Clerbaux et al. 2009; Streets et al. 2013).

Most estimates of air quality from satellite observations have focused on ground-level aerosol mass concentration. However, information on ground-level trace gases, including NO_2, O_3, and CO concentrations, is becoming available. The development of a variety of algorithms to extract physical parameters by accounting for atmospheric radiative transfer and ground-level observations has been integral to the success of modern remote sensing.

7.3 Medium-Resolution Satellite Remote Sensing

Early efforts using medium-resolution satellite remote sensing to assess the distribution of aerosols were accomplished by Fraser et al. (1984) using GOES, by Kaufman et al. (1990) using National Ocean and Atmospheric Administration (NOAA)'s advanced very high resolution radiometer (AVHRR) data, and by Sifakis and Deschamps (1992) using SPOT imagery. Table 7.1 contains an overview of characteristics of the major satellite instruments designed for remote sensing of aerosols and chemically reactive trace gases in the lower troposphere. The particulate mass can be derived from satellite measurements of the radiance of sunlight scattered by Earth's atmosphere. Such a method is most successful where the surface reflection is weak and fairly uniform, as for oceans.

7.3.1 ERS-2

In 1995, the GOME-1 instrument aboard the ERS-2 satellite was launched and information obtained from GOME-1 began to be used for the lower troposphere (Burrows et al. 1999). GOME-1 designed to measure O_3, NO_2, and related species is a nadir-viewing grating spectrometer that measures solar backscatter with broad spectral coverage from 0.230 to 0.790 µm. The spatial resolution of a ground scene is about 40×32 km^2. Global measurements were made by GOME-1 from July 1995 to June 2003, when the tape recorder on ERS-2 failed. Since then, the instrument only has 40% of global coverage using the ERS-2 direct broadcast and a network of receivers until 2011. Several trace gases of troposphere, including NO_2, HCHO, SO_2, and tropospheric O_3, were retrieved globally from GOME (Callies et al. 2000; Martin 2008).

7.3.2 Terra

The Terra satellite was launched by NASA in December 1999. Three instruments onboard, including the MOPITT, the MISR, and the MODIS significantly expanded scientific perspective about the scale of tropospheric pollution. The MOPITT instrument is a nadir-viewing gas correlation radiometer operating in the 4.7 mm band of CO (Drummond 1992). It has a spatial resolution of 22×22 km^2 at nadir with a 29 pixel wide swath and retrieves CO with a target accuracy and precision of 10% (Deeter et al. 2003). The MISR (Diner et al. 1998) and MODIS (Barnes et al. 1998) instruments measure PM and its optical effect and provide unprecedented information about aerosol abundance

TABLE 7.1

Characteristics of Major Satellite Instruments and Chemical Species Detected by these Instruments

				Sensor								
Satellite Platform	ERS-2	Terra			Aqua		Envisat	Aura		Netop-A		GOSAT
Agency	ESA	NASA			NASA		ESA	NASA		ESA/Eumetsat		JAXA
Equator-Crossing Time (LST)	10:30 am	10:30 am			1:30 pm		10:00 am	1:45 pm		9:30 am		1:00 pm
Instrument	GOME	MOPITT	MISR	MODIS	MODIS	AIRS	SCIAMACHY	OMI	TES	GOME-2	IASI	TANSO-FTS
Operational Period	1995–2003	2000–	2000–	2000–	2002–	2002–	2002–2012	2004	2004–	2006–	2006–	2009–
Spatial/Temporal Features												
Global Coverage	3 days	3 days	9 days	1–2 days	1–2 days	Daily	6 days	Daily	2 days	1.5 days	Twice daily	3 days
Spatial Resolution (km)	40–320	22 × 22	17.6 × 17.6	1 × 1	1 × 1	50 × 50	30 × 60	13 × 24	5.3 × 8.5	40 × 80	50 × 50	10 × 10
Spectral Region	UV-Vis	IR	Vis-IR	Vis-IR	Vis-IR	Vis-IR	UV-Vis	UV-Vis	IR	UV-Vis	IR	IR
Spectral Range (μm)	0.24–0.79	2.2,2.3,4.6	0.44–0.87	0.4–14.4	0.4–14.4	3.7–15.4	0.24–2.4	0.27–0.50	3.2–15.4	0.25–0.79	3.6–15.5	0.76–14.3

(Continued)

TABLE 7.1 (Continued)

Characteristics of Major Satellite Instruments and Chemical Species Detected by these Instruments

Satellite Platform					Sensor		
	ERS-2	Terra	Aqua	Envisat	Aura	Netop-A	GOSAT
Agency	ESA	NASA	NASA	ESA	NASA	ESA/ Eumetsat	JAXA
Species Observed							
NO2	x			x	x	x	
SO2	x		x	x	x	x	
CO		x	x	x	x	x	
CH4		x	x	x	x	x	x
NMVOC	x			x	x	x	
PM (AOD)	x	x	x	x	x	x	
NH3					x	x	
CO2				x	x	x	x

and properties at high spatial resolution. The MISR instrument includes nine fixed push-broom cameras that have a capability of pointing at angles varying from +70°, through nadir, to −70°. Each camera has four line-array charge-coupled devices covering spectral bands centered at 0.446, 0.558, 0.672, and 0.867 µm, and having spectral widths of 0.020–0.040 µm, giving a total of 36 channels. The highest spatial sampling is 275 m at all angles. MISR's standard acquisition mode captures full-resolution data in all four nadir-viewing channels and the red-band channels at the other eight angles; the remaining 24 channels provide data at 1.1 km. The retrieval of the MISR instrument provides information on aerosol size, single scattering albedo, and sphericity. The retrievals have been validated with ground-based measurements from AERONET, having a typical accuracy over land of better than 0.05 ± 20%.

The MODIS instrument has 36 channels spanning the spectral from 0.410 to 14.20 µm. The instrument has three spatial resolutions of 250, 500, and 1000 m, depending on the channel. MODIS retrievals aerosol over land from two independent retrievals conducted at 0.470 and 0.660 µm, and subsequently interpolated to 550 µm. Meanwhile, the surface reflectances for the channels at 0.470 and 0.660 µm are estimated from measurements at 2.1 mm using empirical relationships. The retrieved AOD is validated with AERONET and a typical accuracy is between 0.05% and 15%.

7.3.3 Envisat

Envisat was launched on March 1, 2002, and carried the SCIAMACHY (Bovensmann et al. 1999), which has been used for observing a wide range of trace gases in the troposphere. SCIAMACHY measures backscattered solar radiation upwelling from the atmosphere using eight channels with the spectrum over 0.214–1.750 µm at resolution of 0.2–1.4 nm, and two spectral bands around 2.0 and 2.3 µm, having a spectral resolution of 0.2 nm. The typical spatial resolution of SCIAMACHY is 30 × 60 km². Data collection from SCIAMACHY ended when contact with Envisat was lost on April 8, 2012.

7.3.4 Aqua and Aura

Aqua was launched by NASA on May 4, 2002. It carried a second MODIS instrument as well as an atmospheric infrared sounder (AIRS) (Aumann et al. 2003). AIRS is a cross-track scanning grating spectrometer that covers the 3.7–16 µm spectral range with 2378 channels and a 13.5 km nadir field-of-view. CO retrievals are conducted at 4.7 mm with a spatial resolution of 45 km at nadir. Furthermore, AIRS CO measurement agrees with MOPITT CO over land to within 10–15 ppbv without systematic bias, but AIRS CO exhibits a positive bias of 15–20 ppbv relative to MOPITT CO over ocean.

On July 15, 2004, the Aura satellite was launched by NASA to improve our understanding of the changing chemistry of Earth's atmosphere. Two Aura instruments are particularly valuable in an air pollution context: the OMI

(Levelt et al. 2006) and the TES (Beer et al. 2001). OMI instrument is a nadir-viewing imaging spectrometer that uses two-dimensional charge-coupled device detectors to measure the solar radiation backscattered by Earth's atmosphere and surface over 0.270–0.500 μm with a spectral resolution of 0.5 nm (Levelt et al. 2006). The OMI has a spatial resolution of 13×24 km^2 at nadir and a coarser resolution at larger viewing angles. Retrieval algorithms of tropospheric trace gases and their expected uncertainty for OMI are similar to those for GOME-1 and SCIAMACHY, which are described in Section 7.3.3. Furthermore, AOD for both absorption and extinction is being retrieved from OMI. A residual technique and a forward trajectory model are used to retrieve tropospheric O$_3$ columns from OMI. The TES instrument uses a Fourier transform infrared emission spectrometer with high spectral resolution (0.1 cm^{-1}) and a wide spectral range (650–3050 cm^{-1}). The TES nadir has a footprint of 5×8 km^2 and has 71 observations per orbit. Both tropospheric O$_3$ and CO are retrieved by TES.

7.3.5 MetOp

The MetOp-A satellite was launched on October 19, 2006 through a joint initiative of MetOp mission for operational meteorology by ESA and the European Organization for the Exploitation of Meteorological Satellites. MetOp is the first operational meteorological platform that has two instruments dedicated to making tropospheric trace gas measurements: GOME-2 and the infrared atmospheric sounding interferometer. GOME-2 is a nadir-scanning double spectrometer covering the 0.240–0.790 μm wavelength range with a spectral resolution of 0.26–0.51 nm. GOME-2 measures all the species of GOME-1 at a typical resolution of 40×80 km^2. Retrievals of tropospheric trace gases from GOME-2 are similar to these from GOME-1, SCIAMACHY, and OMI. The infrared atmospheric sounding interferometer, which is capable of measuring CO, NH$_3$, CH$_4$, and other species, consists of a Fourier transform spectrometer connected with an imaging system, designed to measure the infrared spectrum emitted by the Earth in the TIR using nadir geometry. The instrument provides spectra coverage from 645 to 2760 cm^{-1} and has high radiometric quality at 0.5 cm^{-1} resolution. The field-of-view of the infrared atmospheric sounding interferometer is sampled by a matrix of 2×2 circular pixels with 12 km diameter each. Measurements are taken every 50 km at nadir with broad horizontal coverage. A nonlinear artificial neural network algorithm is employed for the operational retrievals of tropospheric O$_3$ and CO.

7.3.6 Other Platforms

The Japan Aerospace Exploration Agency (JAXA) and collaborating institutions launched the Greenhouse gases Observing SATellite (GOSAT) on January 23, 2009.

The GOSAT instrument was designed to measure concentrations of CO_2 and CH_4 and their variability over time and space (Kuze et al. 2009). The satellite is equipped with a greenhouse gas observation sensor (TANSO-FTS) and a cloud/aerosol sensor (TANSO-CAI). TANSO-FTS contains detectors for four bands with spectral range of 0.76–14.3 μm and spectral resolution of 0.2 cm^{-1}.

It is worth to mention that Landsat and AVHRR were designed to monitor surface properties at high spatial resolution. However, under some conditions, they can be applied to retrieve AOD with high uncertainty.

7.4 Species Observations

This section introduces the broad principles of satellite detection of atmospheric constituents by focusing on several important emitted species that are observable from space at the present time.

7.4.1 Trace Gases

Retrievals of trace gases through remote sensing include both solar backscatter and TIR emission. Remote sensing by using solar backscatter for trace gas takes advantage of attenuation in the intensity of radiation traversing a medium. The attenuation is usually determined by Beer's law:

$$I_\lambda = I_{\lambda,0}e^{-\sigma_\lambda \rho_s} \tag{7.1}$$

where:
  I_λ is the backscattered intensity observed by a satellite sensor at a specific wavelength λ
  $I_{\lambda,0}$ is the backscattered intensity that would be observed in the absence of absorption
  σ_λ is the absorption cross section of the trace gas
  ρ_s is the trace gas abundance over the atmospheric path length

Trace gas retrievals using solar backscatter estimate spectral variation in σ to infer ρ (Chance 2006). Retrievals comprise a spectral fit to determine atmospheric abundance over the radiation path, and a radiative transfer calculation to determine the path of radiation through the atmosphere. Usually, the radiative transfer calculation is of particular importance at ultraviolet and short visible wavelengths. However, the land surface reflectivity is typically less than 5% at these wavelengths (Herman and Celarier 1997) and molecular scattering is a major contributor to backscattered radiation. The instrument sensitivity to trace gases in the lower troposphere is related the intensity of reflectivity. Consequently, scattering by clouds usually enhances the instrument sensitivity to the trace gas above the cloud and decreases the instrument sensitivity to the trace gas below the cloud. Similarly, aerosols can either enhance or reduce the instrument sensitivity depending on their single-scattering albedo and vertical distribution.

Optically thin cases that meet the condition of $\sigma_\lambda \rho_s < 1$ allow separation of the solar backscatter retrieval into an independent spectral fit and air mass factor (AMF) calculation (Martin 2008). An AMF is determined by the ratio of ρ_s to the vertical column ρ_v.

$$AMF = \frac{\rho_s}{\rho_v} \qquad (7.2)$$

The AMF calculation can decouple the vertical dependence of the sensitivity to the trace gas from the shape of the trace gas vertical profile. A radiative transfer model is used to calculate the sensitivity of backscattered radiation at the top of the atmosphere to the vertically resolved trace gas concentration in the atmosphere. The local shape of the trace gas vertical profile is generally calculated with an atmospheric chemistry model.

For optically thick ($\sigma_\lambda \rho_s > 1$) condition, Equation 7.2 is not appropriate. Both the spectral fit and radiative transfer calculation are needed. The log of the optical thickness τ_λ of several gases affecting air quality for nominal atmospheric concentrations, where $\tau_\lambda = \sigma_\lambda \rho_v$ verses wavelength indicates that all trace gases are optically thin at wavelengths longer than 320 nm (Martin 2008). O_3 is the dominant absorber at wavelengths shorter than 350 nm. The SO_2 spectrum exhibits similar spectral structure, but its optical depth is smaller than that for O_3.

The ability of nadir-viewing satellite instruments to detect trace gases and aerosols in the atmosphere depends on several factors, including the surface reflectivity or emissivity, clouds, the viewing geometry, and the retrieval wavelength. Additionally, the detection of trace gases depends on its vertical profile and, for solar backscatter, on aerosols. The instrument sensitivity to trace gases in the middle and upper troposphere is typically near perfect. However, atmospheric scattering and emission cause part of the measured intensity not passing through the boundary layer, and therefore decrease the instrument sensitivity to trace gases and ultraviolet-absorbing aerosols near the ground. The decrease in sensitivity in the boundary layer is particularly pronounced at ultraviolet wavelengths (Klenk et al. 1982) or in the TIR due to reduced thermal contrast between the atmosphere and surface (Pan et al. 1995). However, the spatial distribution of trace gases and aerosols has implications for their retrieval. But the stratospheric column and the vertical profile in the troposphere are two major issues. High variability in the stratospheric O_3 column constrains the discrimination of the tropospheric O_3 column at middle and high latitudes (Schoeberl et al. 2007). In contrast, relatively weak zonal variability in the stratospheric NO_2 column enables separation of the stratospheric and tropospheric columns. Normalized annual mean vertical profiles of trace gases and aerosol extinction calculated with a global chemical transport model (Bey et al. 2001) over the United States and southern Canada suggests that normalizations of these species by their column abundances facilitate comparison of the relative vertical profile (Martin 2008). For instance, concentrations of tropospheric NO_2 and SO_2 are enhanced strongly in the boundary layer due to strong surface sources, short lifetimes, and increase in the NO/NO_2 ratio with altitude that is driven by the temperature dependence of the $NO + O_3$ reaction. Consequently, column observations of NO_2 contain large contributions from the boundary layer. HCHO columns and optical thickness of aerosol are also strongly affected by boundary layer enhancements in both HCHO concentration and aerosol extinction. In contrast, weak vertical variation in the number density of O_3 and CO suggests that vertical profile information in both species is often necessary to extract a boundary layer signal. Boundary layer O_3 constitutes only a small fraction (usually less than 2%) of the total O_3 column.

7.4.2 Aerosol Remote Sensing

The principle to measure aerosols in the atmosphere is based on the optical atmospheric effects. The signal measured by a remote sensor is influenced by the atmosphere in two ways: radiometrically and geometrically. They can modify the signal's intensity through scattering or absorbing processes and its direction by refraction. Most measurements of aerosol using satellite imagery focused on radiometric variations. The quantity designating the atmospheric effect to these alterations is the atmospheric optical depth (AOD) or thickness. AOD is a dimensionless number that expresses the total influence of both molecular and particulate atmospheric constituents on a radiation at a wavelength λ, which traverses the atmosphere vertically.

AOD is expressed as the integration of aerosol extinction coefficients $[\sigma_{ext}(z, \lambda)]$ along the vertical atmospheric column from the ground to the top of the atmosphere:

$$AOD = \int_0^\infty \sigma_{ext}(z,\lambda)dz \qquad (7.3)$$

Total AOD can be broken down into the sum of the partial optical depths including optical depth due to molecular absorption, optical depth due to particular absorption, optical depth due to molecular scattering, and optical depth due to particulate scattering.

However, the satellite-derived AOD is measured from a column AOD in ambient conditions. The $PM_{2.5}$ mass only represents the mass of dry particles near the surface. Factors that are directly related to the mass of $PM_{2.5}$, such as mass extinction efficiency, hygroscopic growth factor, which is a function of relative humidity, and effective scale height, which is mainly determined by the vertical distribution of aerosols, cannot be derived from AOD. Furthermore, the fluctuations of aerosol mass concentration profile were usually observed and could introduce uncertainties in the relationship between satellite-derived AOD and $PM_{2.5}$ mass (Coutant et al. 2003; Wang and Christopher 2003). Therefore, these parameters are usually derived by using measurements and models.

7.5 Assessments of Air Quality in Urban Areas

This section describes comprehensive applications of satellite detection of atmospheric constituents, focusing on four important emitted species that are commonly observable from space at the present time. The applications of retrieved aerosols and trace gases in several urban regions are also presented.

7.5.1 Assessment of $PM_{2.5}$ Distribution

A simple, yet effective approach has been developed (Liu et al. 2004a, 2004b) to connect the spatial and seasonal variation of factors that are essential in retrieving AOD and $PM_{2.5}$ concentration from satellite imagery. The method applies local scaling factors simulated from a global atmospheric chemistry model to reduce the uncertainty in estimated $PM_{2.5}$ concentrations. The formula used to calculate the concentration is expressed as follows:

$$\text{Concentration of PM}_{2.5} = \left(\frac{\text{Model surface aerosol concentration}}{\text{Model AOD}} \right)$$
$$\times \text{Retrieved AOD}$$

In this equation, the model surface aerosol concentration is simulated from the GEOS-CHEM model with GOCART dust and sea salt data. The GEOS-CHEM model is a global three-dimensional tropospheric chemistry and transport model driven by assimilated meteorological observations (http://www.acmg.seas.hardvard,edu/geos/index.html). Model AOD is calculated using aerosol dry mass concentrations with consideration of impacts of increased relative humidity on particle growth by applying different hydroscopic growth factors to all hydrophilic species using relative humidity conditions (Bey et al. 2001; Martin et al. 2003; Park et al. 2003). The retrieved AOD is derived from using MISR boarded on the NASA EOS satellite Terra and launched in 1999. The $PM_{2.5}$ concentration derived from this algorithm has a relative error of approximately 20% (Liu et al. 2005).

An explicit method has been implemented to formulate the relationship between AOD and $PM_{2.5}$ in order to isolate the parameters involved in aerosol mass estimate (Donkelaar et al. 2006). AOD and total column aerosol mass loading W are determined by

$$W = \frac{4}{3}\left(\frac{\rho r_{\text{eff}} \text{AOD}}{Qe}\right) \quad (7.5)$$

where:

ρ is the aerosol mass density at ambient relative humidity

r_{eff} is the column averaged effective radius (defined as the ratio of the third to second moment of an aerosol size distribution at ambient relative humidity)

Qe is the column averaged extinction efficiency

By including the relative vertical profile of aerosol extinction, the aerosol mass concentration $M_{\Delta z}$ between the ground and altitude Δz can be expressed as

$$M_{\Delta z} = \frac{4}{3}\left(\frac{\rho_{\Delta z} r_{\Delta z,\text{eff}} f_{\Delta z}}{Q_{\Delta z,e}\Delta z}\right)\text{AOD} \quad (7.6)$$

where:

f represents the fractional optical thickness below altitude Δz

All parameters in the bracketed expression refer to representative values below altitude Δz. Accounting for aerosol hygroscopicity and assuming spherical aerosols and uniform aerosol properties between the surface and altitude Δz, the dry mass of $PM_{2.5}$ at the surface can be formulated as

$$M_{2.5,d,\Delta z} = \left[\frac{4}{3}\left(\frac{R_{2.5,d,\Delta z,\text{eff}}}{R_{2.5,\Delta z,\text{eff}}}\right)^3\left(\frac{\rho_{2.5,d,\Delta z}R_{2.5,\Delta z,\text{eff}}f_{2.5,\Delta z}}{Q_{2.5,e,\Delta z}\Delta z}\right)\right]\text{AOD} \quad (7.7)$$

where:

the subscript d indicates dry conditions and the subscript 2.5 denotes aerosols smaller than 2.5 μm in diameter

$R_{2.5,d,\Delta z,\text{eff}}$ represents the fine dry effective radius

$f_{2.5,\Delta z}$ represents the ratio of fine AOD below altitude Δz to the total AOD

$M_{2.5,d,\Delta z}$ is the total fine dry aerosol mass between the surface and altitude Δz

All these parameters except AOD can be calculated from a chemical transport model (GEOS-CHEM). The GEOS-CHEM is driven by assimilated meteorological data from the Goddard Earth Observing System (GEOS-3) at the NASA Global Modeling Assimilation Office to calculate these values. Meteorological fields used in the model include surface properties, humidity, temperature, winds, cloud properties, heat flux, and precipitation. This approach has been implemented with AOD measured by the MODIS and MISR instruments.

The ground-level $PM_{2.5}$ measured from the Canadian National Air Pollution Surveillance network and the U.S. Air Quality System was also used to compared with model derived $PM_{2.5}$. The spatial variation in annual mean of the model derived $PM_{2.5}$ shows significant agreement with surface measurement when using AOD from MODIS and MISR. The daily variation in remotely sensed $PM_{2.5}$ was more consistent with surface measurements in North America. Furthermore, simulation of parameters in GEOS-CHEM improves the spatial correlation of remote and surface $PM_{2.5}$ from 0.36–0.37 to 0.58–0.69. In contrast, daily variation in AOD played the major role in accurately representing daily variation in remotely sensed $PM_{2.5}$. Daily variation in parameters such as the relative vertical profile of aerosol extinction or the effective radius was insignificant.

Spatial and temporal variations of surface $PM_{2.5}$ concentration simulated from satellite remote sensing and GEOS-CHEM across the conterminous United States and part of Canada displays a seasonal maximum during summer, corresponding to higher sulfate levels produced by increased SO_2 oxidation rates (Donkelaar et al. 2006). Elevated in $PM_{2.5}$ concentration in California is associated with high emissions of aerosol precursors in urban areas. However, high values of surface $PM_{2.5}$ in the southwestern United States could be caused by an artifact in the retrieval.

A simple correction method has been implemented to estimate concentrations of PM_{10} and $PM_{2.5}$ using MODIS imagery (Wang et al. 2010). This method uses vertical and relative humidity correction on the surface measurements and AOD. The corrective models between AOD and PM have been implemented with use of MODIS 1 km AOD data to estimate the PM concentrations in the Beijing area, China. The distribution of PM_{10} and $PM_{2.5}$ in Beijing urban area in 1 km resolution suggests that concentrations of PM in central urban areas are apparently larger than these in noncentral urban area. The highest value usually emerges around the center of Beijing, where the more urbanized districts and several large economic-technological or industrial development zones are located. This area only counts about 8.31% of the city but dwellings about 61.98% of the population. Apparently, the PM concentrations reduce progressively with distance away from the central part. In the mountain and rural areas, where there are less residential or industrial activities, the PM concentrations become much lower. The minimum appears in the western mountains where the average altitude is above 600 m.

7.5.2 Nitrogen Dioxide

Nitrogen oxides (NO_x) are one of important classes of atmospheric species that have a number of environmental implementations, including the formation of tropospheric ozone and aerosols, acidification, eutrophication, and human health effects. NO_x is primarily produced by man-made sources through combustion processes. However, emission factors vary according to the fuel type and combustion conditions, resulting in large uncertainty in

NO_x emissions. There is potential for measuring NO_x emission in a large area using satellite instruments. Tropospheric NO_2 columns are closely related to surface NO_x emissions due to the short NO_x lifetime and an increasing NO/NO_2 ratio with altitude.

Algorithms have been developed to measure NO_x. For example, the OMI NO_2 algorithm employs the OMI-measured earthshine radiances and solar irradiances to calculate slant column NO densities, processed with an initial AMF to get initial estimates of the vertical column densities (Bucsela et al. 2006). A whole-day field of initial vertical column density is used to develop smooth approximation of the global stratospheric field and identify small-scale regions that have significantly elevated NO_2, where the vertical column density is recalculated using an AMF for the tropospheric component. The algorithm contains a number of parameters that specify how the retrieval is accomplished.

Observations from the OMI instrument have been directly used to examine four urban regions in California, United States (Russell et al. 2010). In California, efforts to reduce O_3 and aerosol by reducing their chemical precursors (NO_x, volatile organics and directly emitted particles) have resulted in significant improvements in air quality (Russell et al. 2010). However, concentrations of ozone and aerosol still remain higher than the regulatory limits over the most populous parts of the state. Observations from the OMI were processed to outline the spatial and temporal variability of NO_2 concentrations in four metropolitan regions of California during the years 2005–2008. Several major urban areas are included in these regions: Sacramento, San Francisco, Los Angeles, and Fresno-Bakersfield. Figure 7.1 illustrates geographic locations of these regions. OMI achieves complete global coverage each day with a repeat cycle of 16 days; thus for a given location, the NO_2 column is gathered by a different pixel within the 60 across track positions each day. Consequently, data are collected at a different spatial resolutions and with pixel centers at different locations each day. The data were processed using an area weighted averaging, binning to $0.025 \times 0.025°$ by including the center 20 pixels from each swath to limit the measurements to those with a pixel size less than 15×27 km^2, and eliminate pixels at the outer edges of the swath that have dimensions as large as 24×128 km^2, resulting in a much higher spatial resolution than maps that include the larger pixel size measurements.

Different thresholds were selected to highlight areas with relatively higher NO_x concentrations depending on locations. Areas where the NO_2 column in summer 2005 exceeded 6×10^{15} molecules/cm^2 were identified in the Sacramento, San Joaquin Valley, and South Coast regions. In 2005, urban plume from the San Francisco Bay area extended east to blend with the plumes from other cities. To define the region of interest for San Francisco, measurements from summer 2008 were used with the threshold of 4.6×10^{15} molecules/cm^2 for the 2008 measurements that is comparable to using the 6×10^{15} molecules/cm^2 threshold for 2005 data. The summer months are focused because the NO_x photochemical lifetime is shorter (~4 h at noon). Relatively high NO_x were observed in Sacramento County, the San Francisco Bay area air basin, the

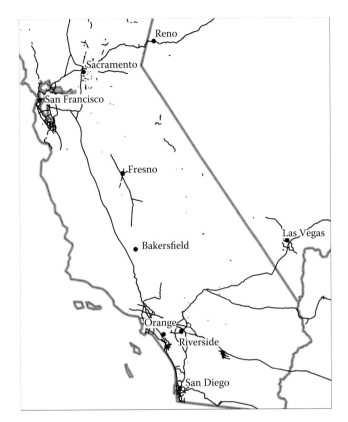

FIGURE 7.1
Locations of four metropolitan regions in California and major roads. This region is the study area for estimating O_3 concentration conducted by Russell et al. Major roads are also displayed. (Data from Russell, A.R. et al., *Environmental Science & Technology*, 44, 3608–3615, 2010.)

San Joaquin Valley air basin, and the South Coast air basin. Observations showed that elevated concentrations of NO_2 were detected in and around several highly populated areas, including the major cities of California such as Sacramento, San Francisco, Fresno, Bakersfield, Los Angeles, and San Diego in California; Reno and Las Vegas in Nevada; and Tijuana in Mexico. In all of these regions, the NO_2 columns were higher on weekdays than on weekends, suggesting weekday emissions were highly related to transportation activities. Outside of these population hotspots, no significant day-of-week variability in NO_2 column was detected in rural regions.

OMI observations can be averaged to reveal NO_2 concentration in a large area. A three-year average of tropospheric NO_2 columns retrieved from OMI for the conterminous United States suggested that distinct enhancements were evident over major urban and industrial regions in both western and eastern coasts. The high degree of spatial heterogeneity provides empirical evidence that the dominant signals are from the boundary layer. For example, *in situ* aircraft measurements of NO_2 indicate that 75% of the NO_2 column is in the lowest 1500 m over Houston (Martin et al. 2004). Tropospheric NO_2 columns are closely related to surface NO_2 concentrations.

7.5.3 Ozone

Tropospheric ozone is produced by complex photochemical reactions involving nitrogen oxides (NO_x, which includes NO and NO_2) and hydrocarbons emitted from both anthropogenic and natural sources. During daytime, the photochemical oxidation of VOCs helps cycling of NO to NO_2, and the later facilitates photochemical dissociation of NO_2 through the following reaction:

$$NO_2 + hv \rightarrow NO + NO \tag{7.8}$$

The formation of ozone from the combination of O and molecular O_2 occurs through the reaction

$$O + O_2 \rightarrow O_3 \tag{7.9}$$

Consequently, the abundance of NO_x near the surface level leads to removal of O_3 through the reaction of NO and O_3:

$$NO + O_3 \rightarrow NO_2 + O_2 \tag{7.10}$$

The emission of NO from automobile engines in the urban area and the subsequent chemical reaction described in the equation is responsible for the partial destruction of surface ozone, especially during nighttime. The formation of O_3 is also influenced strongly by climatological factors including air temperature, moisture, and solar radiation.

Retrieval of atmosphere boundary layer O_3 from satellite remote sensing is still a challenge. Current O_3 retrievals show low sensitivity in the boundary layer due to molecular scattering in the ultraviolet and surface emission in the TIR (Martin 2008). Furthermore, boundary layer O_3 only counts a small fraction of the total column.

Currently, the Aura satellite has been used for direct measurements and retrievals of tropospheric ozone. The measurement is made from the on-board instrument in different spectral regions: the TES in the TIR (Beer 2006) and the OMI in the ultraviolet (Levelt et al. 2006; Liu et al. 2010). Consistency between TES and OMI measurements has been demonstrated (Zhang et al. 2010). However, the detection of urban pollution signatures from space is restricted by several factors, such as the lifetime of the species, the sensitivity of the measurements to the lower troposphere, and the vertical profile of the species. Detection of ozone signature for cities using the tropospheric column data is more difficult because only about 10% of the total ozone column resides in the troposphere.

NASA's TOMS instrument uses a different retrieval algorithm from OMI with only six wavelength bands, from which the ozone column can be measured very accurately. TOMS has the advantage that it has a fairly small ground-pixel size (50×50 km^2) in combination with a daily global coverage. Datasets from TOMS have been used to detect plumes of tropospheric ozone from several large and polluted cities around the world (Kar et al. 2010). The troposphere ozone residual distributions and the troposphere column ozone data in the vicinity of Beijing (August 2005), New York (July 2005), São Paulo (October 2005), and Mexico City (February 2005) were estimated from TOMS measurements. The enhanced tropospheric ozone observed simultaneously in both the datasets suggests that the plumes may be associated with the cities and not artifacts of the retrieval. The 850 hPa wind trajectories were added in these assessments for Beijing, New York, and São Paulo and 700 hPa for Mexico City. The weak wind trend in Beijing enhanced ozone photochemistry. The orientation of the ozone plume was consistent with the west wind and extends over the Atlantic Ocean in the New York area. Ozone plumes could be extended northeast to Massachusetts and Maine originating from the New York plume in summer. In addition, the typical lifetime of ozone is about two to five days in the troposphere in the boundary layer and is about 90 days in the middle troposphere. The ozone plums produced in urban areas could be transformed to hundreds of kilometers away and generate impacts in much large areas.

7.6 Urban Land Cover and Air Quality

This section illustrate the use of spatial and temporal urban development information derived from satellite remote sensing data to assess the relationship between urban land cover and land use and air quality in the Las Vegas Valley, Nevada, United States. Air quality observations of O_3, NO_x, and $PM_{2.5}$ concentration were analyzed for the region.

The Las Vegas Valley is characterized by an arid climate and desert landscape. Dry air masses are the dominant influence on the valley region and produce clear and partly cloudy skies most of the year. Ozone was one of the air quality concerns in the Las Vegas Valley, which was designated by the US EPA as a nonattainment area for failing to meet a new eight-hour standard. $PM_{2.5}$ was also a concern because of the desert environment in the region. Surface measurement records for O_3, NO_x, $PM_{2.5}$, and temperature were acquired from several Clark County air quality monitoring sites.

Urban development in the Las Vegas region was determined using the percent anthropogenic impervious surface that was quantified using regression tree models. Details about how impervious surface and urban land cover was estimated has been illustrated in Chapter 3 and in a previous study (Xian 2007). Landsat ETM+ imagery in 2002 and high-resolution orthoimages were used to quantify urban land cover for the Las Vegas. To evaluate urban land cover influences on air pollutant variations, both means of impervious surface area (ISA) and normalized difference vegetation index (NDVI), retrieved from satellite imagery, were calculated within both 1 km by 1 km and 2×2 km^2 grids around each of the selected monitoring sites in the valley. The correlation analyses between urban land cover and air pollutant concentrations were conducted. These included correlation coefficients between the pollutant concentrations measured in different sites and associated ISA, as well as NDVI averaged over the two size grids. Records of 2002 ozone levels observed from eight sites and NO$_x$ levels observed from four sites were used. Records of PM$_{2.5}$ levels from four sites CR, ES, JD, and PV were also used for correlation calculations. Figure 7.2 illustrates both impervious surface and NDVI greenness derived from 2002 Landsat imagery. Spatial locations of air quality monitoring sites in the valley are also shown in the figure. Urban land use expanded in all directions across the valley, with the highest percent ISA categories extended from the downtown and the Las Vegas strip to the both southeast and northwest portions of Las Vegas. Most built-up areas belonged to medium-to-high imperviousness densities. The urban landscape in the area was characterized as developed urban land mixed with planned vegetation canopy throughout the metropolitan area. Despite the arid climate that characterizes the environs of the valley, the green areas visible on the NDVI image reveals a relatively dense vegetation canopy existing throughout much of the urban area. Strong greenness intensities were associated with most residential areas.

The spatial and temporal O$_3$ distribution data from eight monitoring sites that provide coverage for the Las Vegas Valley were analyzed. To evaluate

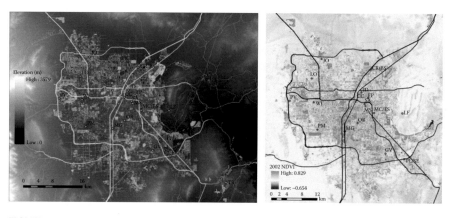

FIGURE 7.2
Urban impervious surface (left) and NDVI (right) in 2002 in the Las Vegas Valley. Air quality monitoring sites are also displayed.

the spatial variation of O_3 in the valley, the monitoring sites were grouped by geographic region: southern sites included PM, PL, and JO; northern sites included JD, CC, PV, WJ, and LO; western sites included LO, PV, WJ, PM, and JO; and eastern sites included JD, CC, and PL. Monthly average total O_3 distributions together with daytime and nighttime monthly averages obtained from records between 1996 and 2004 for the valley are shown in Figure 7.3. The total O_3 concentrations reached their peaks during May and June (Figure 7.3a), and concentrations were consistently higher on the west side of the valley as opposed to the east side (Figure 7.3b). A small seasonal variation was evident in the north–south distribution of ozone, where higher concentrations of O_3 occurred in the southern part of the valley during summer, while higher concentrations of O_3 appeared in the northern part of the valley during winter. These relationships also hold true for daytime and nighttime ozone distributions. During daytime, the photochemical oxidation of VOCs helps in cycling of NO to NO_2, and the later facilitates photochemical dissociation of NO_2. The formation of ozone from the combination of O and molecular O_2 then occurs, resulting in daytime O_3 levels being much higher than those at nighttime. Consequently, the abundance of NO_x near the surface level leads to removal of O_3. The emission of NO from automobile engines in the urban area and the subsequent chemical reaction are responsible for the

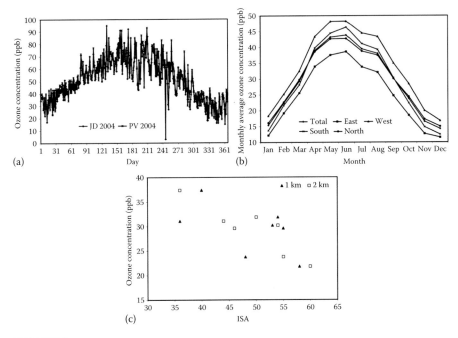

(a)

(b)

(c)

FIGURE 7.3
Daily maximum surface O_3 in 2004 at monitoring sites JD and PV (a), multi-year monthly average of O_3 distributions of sites in four directions and all site (b), and correlation between 2002 annual average ozone concentration and percent imperviousness with 1 and 2 km grids (c).

partial destruction of surface ozone, especially during nighttime. An apparent diurnal variation of NO_x concentration levels showed that the nighttime NO_x level was higher than that in daytime.

The spatial distributions of O_3, NO_x, and $PM_{2.5}$ concentrations had strong heterogeneous features in the Las Vegas Valley. Some of the major pollutants in the valley were transported from the western side of the valley associated with prevailing winds and some of them were released by local anthropogenic sources. The distributions and life cycles of these pollutants highly depended upon regional and local atmospheric conditions, as well as local land cover and land use characteristics. Most urban land use in the valley was categorized as residential and commercial development according to ISA estimation illustrated in Figure 7.2. One effect of the increasing of anthropogenic impervious surface was to enhance the urban heat island effect and elevate local air temperatures. This has led to increases in VOC emission from urban trees, vegetation, and anthropogenic sources. Urban land cover and land use conditions displayed in Figure 7.2 suggested that most monitoring sites were located in areas where high-intensity impervious surface areas were mixed with vegetation coverage. High correlations between ozone concentrations and mean percent ISA over 1 km and 2 km grids were found. The correlation coefficient of daytime ozone with percent ISA over a 1 km grid was −0.64 and over a 2 km grid was −0.77, which was the highest value for all ozone concentration categories. Furthermore, the correlation of ozone concentration for a 2 km grid was significant at 95% confidence level, whereas the correlation for a 1 km grid was significant at 90% confidence level. Figure 7.3c illustrates the correlation between annual average ozone concentration and percent ISA averaged over two grid spaces. Furthermore, the correlation analysis for ozone and NDVI suggests that the ozone concentrations had almost no correlation with NDVI, although daytime ozone concentrations showed relative weak positive correlations with NDVI for both grid space cases.

7.7 Summary

In this chapter, the major satellite instruments used for detecting and retrieving trace gases and aerosols for air quality applications have been introduced. Algorithms of retrieval of air quality using satellite remote sensing of the composition of the boundary layer over land, which is of direct relevance for surface air quality due to rapid vertical mixing during the day, is specifically focused.

As described in Section 7.1, urban areas are major sources for several pollutants including aerosols, NO_x, and O_3. Air quality is one of major health issues in many urban areas. With consideration of spatial and temporal features of pollutant, one of the major health effect assessments is to quantify pollution gradients within city extents. Surface observational networks usually monitor air quality in selected sites. The assessment of air quality in a metropolitan extent requires information about spatial distributions as well as temporal variations of pollutants. Many medium-resolution satellites provide different wavelength regions that are sensitive to different vertical regions of the atmosphere. The feature of variation in sensitivity can be exploited to better discriminate the aerosols and trace gases in boundary layer from the free troposphere.

The current spatial resolution of satellite observations is insufficient to be used to assess air quality in intra-urban scales. Higher-spatial-resolution measurements will be the ultimate solution for determining the spatial distributions for individual species. However, aerosol and trace gas retrievals can also be achieved using external information about the species itself and local information on the species profile and on aerosol properties. Numerical models that can accurately represent *in situ* measurements remain an important source of the profile information. The use of space-based measurements from satellite instruments coupled with additional information from a global chemical transport model is an optimal solution for assessing air quality in urban areas. Furthermore, satellite observations can be directly used to detect urban signatures of trace gases such as ozone and NO_x.

The analysis of urban land cover and air quality suggests that pollutant concentrations are influenced by surface landscape conditions. Percent ISA has a negative correlation with ozone concentration and positive correlations with other pollutants. The negative correlation found between ozone concentrations and percent impervious surface suggests that high ozone concentrations are more closely associated with the medium-to-low density urban areas of Las Vegas. Urban vegetation canopy has a locally positive effect by reducing ozone in urban areas. Furthermore, an apparent negative correlation between urban vegetation and fine PM concentration was also found in the region. The influencing range of urban land use and land cover (LULC) varies with different pollutants. The assessment suggests the apparent local influence of urban development density on air pollutant distributions.

8

Air Quality in Urban Areas—Global Aspects

8.1 Introduction

Urban areas respond to, and are responsible for, changes in biochemical and climate cycles (Grimm et al. 2008). Many urban areas in the world have been undergoing persistent expansions in urban land use, transportation system, and automobile usage. Urban centers that provide habitats for human beings are also point sources of CO_2 and other greenhouse gases, as well as trace gases such as NO, NO_2, O_3, and other organic acids.

Urban area is an important source region of a variety of natural and anthropogenic aerosols. Waste and pollutants generated in cities could enter in atmosphere or water and further could be transported from local to global and the extent of influence depends on the path by which materials are carried away from their sources. For example, particulate matter (PM) (or aerosols) is one of the major pollutants that affects air quality in urban and even rural areas of the world. PM is a complex mixture of solid and liquid particles that vary in size and composition, and remain suspended in the air. Air pollutions generated in many megacities have produced environmental problems in the past few decades because of its hazardous effects on human health and its potential impact on local, regional, and even global climate.

Pollutants are removed from the atmosphere through processes of gravitational settling, dry deposition, and wet scavenging. These processes usually produce lifetimes of hours to days in the atmospheric boundary layer and of weeks in the upper troposphere (Uno et al. 2009; Yu et al. 2013b). Therefore, aerosols emitted from one city or many cities in one continent are often transported to other regions or long distances to another continent, in particular when aerosols are pumped out of the boundary layer. Pieces of evidence provided by long-term surface monitoring networks, *in situ* measurements from intensive field campaigns, and especially satellite observation, have shown intercontinental and even hemispheric transport of aerosols. Because of the long-range transport, pollutants emitted or formed in urban areas in one region can create significant environmental impacts in downwind regions or even continents. Sometimes, impacts of intercontinental transport on regional air quality and climate change could be large. Aerosols affect local and regional weather and climate by scattering and absorbing solar radiation and by modifying cloud properties, amount, and evolution (Kaufman et al. 2005a). The change in absorption of solar radiation by particles can alter the atmospheric stability structure and reduce the surface flux, which can even change atmospheric circulations.

There is mounting evidence for intercontinental and even hemispheric transport, provided by long-term surface monitoring networks, *in situ* measurements from intensive field campaigns, and especially satellite observation, supported by model simulations. Large-scale dust storms and associated dust clouds from Asia, for example, were transported in the subsequent days over an extraordinarily long distance along the westerlies in the upper troposphere (Uno et al. 2009). The 3D transport structures of the dust across the Pacific were retrieved by the integrated analysis of the NASA's cloud-aerosol lidar and infrared pathfinder satellite observations (CALIOP) satellite measurements and atmospheric transport models. The 3D view of the hemispherical journey of strong dust event, originated from the Taklimakan Desert during May 09–22, 2007, was simulated by CALIOP for vertical distributions of depolarization ratio, extinction coefficient, and a global aerosol transport model for horizontal distribution of dust extinction coefficient. The extended analysis allows a more complete view of the dust generation, transport and deposition, and enables investigations of the dust contribution to the background aerosol.

It is necessary to routinely monitor the long-range transport of air pollution on a global basis. Satellite remote sensing can provide observations for aerosols and other trace gases with global or continental coverage over decadal-scale. Satellite remote sensing has been used for tracing gases and aerosols for the large-scale air pollutant analysis since the 1970s. The early applications of satellite remote sensing of aerosol used the advanced very high resolution radiometer (AVHRR), Landsat, and Geostationary Operational Environmental Satellite (GOES) instruments to trace desert particles across ocean (Fraser 1976; Carlson and Wendling 1977; Mekler et al. 1977). Moderate-resolution imaging spectroradiometer (MODIS) data acquired after 2000 provide more regional and continental view of aerosol generation and transportation.

Figure 8.1 illustrates global distribution of aerosol observed for aerosol optical depth (Figure 8.1a) and size (Figure 8.1b) from MODIS imagery. These maps show average monthly aerosol amounts around the world based on observations from the MODIS on NASA's Terra satellite. Satellite measurements of aerosols, measured as aerosol optical thickness, are based on the fact that the particles change the way the atmosphere reflects and absorbs visible and infrared light. An optical thickness of less than 0.1 (palest yellow) suggests a crystal clear sky with maximum visibility, whereas a value of 1 (reddish brown) represents very hazy conditions. High aerosol amounts are connected with different process in different places and times of year. For

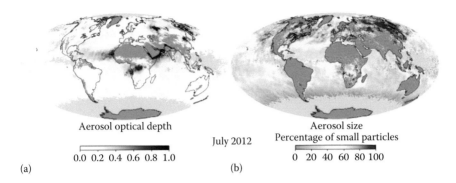

FIGURE 8.1
Global maps of aerosol optical depth (a) and aerosol size (b) developed from NASA Earth Observatory. These maps are based on observations from the moderate-resolution imaging spectroradiometer (MODIS) on NASA's Terra satellite. (Data from NASA Earth Observatory, http://earthobservatory.nasa.gov/GlobalMaps.)

instance, high aerosol amounts could appear over South America from July through September due to land clearing and agricultural fires that are widespread across the Amazon Basin and Cerrado regions during the dry season. A similar seasonal pattern of aerosol distribution can be seen in Central America (March–May), central and southern Africa (June–September), and Southeast Asia (January–April). However, in many other events, aerosol amounts rise dramatically around the Arabian Peninsula and nearby oceans due to dust storms from May to August each year. Increased aerosol amounts placed at the foothills of the Himalayas Mountains in northern India in some months, and linger over eastern China for much of the year. These elevated aerosol amounts are mainly contributed by human-produced air pollution from these regions.

Figure 8.1b displays spatial variations of aerosol size. Aerosol particles of natural origin (such as windblown dust) tend to have a larger radius than human-produced aerosols such as particle pollution. This false-color map shows where there are natural aerosols, human pollution, or a mixture of both on a monthly basis. Green areas represent aerosol plumes dominated by larger particles, while red areas are aerosol plumes dominated by small particles. Yellow areas represent plumes, in which large and small aerosol particles are intermingling, and gray shows no data is collected. The aerosol size distribution also suggests that in the planet's most southerly latitudes, nearly all the aerosols are large, while in the high northern latitudes, smaller aerosols are very abundant. However, in the Southern Hemisphere, where most areas are covered by ocean, the largest source of aerosols is natural sea salts. In contrast, land is concentrated in the Northern Hemisphere and the amount of small aerosols from fires and human activities is greater there than in the Southern Hemisphere. Over most terrestrial regions, patches of large-radius aerosols appear over deserts and arid regions, most prominently, the Sahara

Desert in northern Africa and the Arabian Peninsula, where dust storms are common. Meanwhile, areas where human-triggered or natural fire activity is common such as land-clearing fires in the Amazon from August–October, or lightning-triggered fires in the forests of northern Canada in Northern Hemisphere summer, smaller aerosols are apparent. Human-produced pollution is largely responsible for the areas of small aerosols over many developed areas, including the eastern United States and Europe, especially in the summer season.

In this chapter, we review how the satellite remote sensing is applied to monitor aerosol cross-continental transport. Section 8.2 describes satellite systems that are currently used for monitoring large-scale aerosol transport. Section 8.3 depicts methods used to retrieve aerosol anthropogenic component. Section 8.4 presents applications of satellite retrievals for a large-scale aerosol transport monitoring. Section 8.5 is the summary of the chapter.

8.2 Satellite Systems for Global Air Quality Assessment

Several fundamental requirements need to be met for using reliable observations to assess aerosol intercontinental transport. First, measurements should have sufficient spatial and temporal coverage to allow for tracking the aerosol transport over a global or continental scale with a daily or even hourly frequency. Second, observational data can be used to derive highly accurate, quantitative information about aerosol horizontal and vertical distributions, such as mass concentration, mass loading, aerosol optical depth (AOD), and vertical profiles of aerosol extinction. Third, observational data can be used to distinguish aerosol types, which are essential for identifying aerosol intercontinental transport from the measured bulk of particles. Aerosol type information is required to convert aerosol optical measurements (e.g., AOD and extinction or backscatter coefficient) to aerosol mass loading or concentration. Different types of aerosol also tend to be transported in different paths and apply different impacts on climate and human health.

Satellite measurements, especially those having large-scale spatial coverage and complemented by regional and local aerosol characterizations from surface stations and field campaigns, can meet these fundamental requirements. Regional and local aerosol characterizations from surface stations and field campaigns can provide detailed local measurements of surface concentrations to improve algorithms used to characterize the distribution aerosols. However, even long-term surface networks, such as the Aerosol Robotic Network (AERONET) (Holben et al. 1998), are substantially inadequate in geographical coverage, particularly over oceans where intercontinental transport occurs (Yu et al. 2013a). Satellite remote sensing can extend field campaigns and surface networks by expanding temporal and spatial scales because of the intrinsic advantage of daily or even hourly measurements with global or continental coverage over decadal durations.

Early uses of satellites to characterize aerosol intercontinental transport were just to provide imagery of major aerosol plumes across different continents. Substantial progress has been made in satellite remote sensing of aerosols and their distributions in recent decades (Yu et al. 2009; Streets et al. 2013). Satellite measurements are currently providing global land and ocean measurements of AOD with much better quality. Global measurements of aerosol vertical distributions in cloud-free atmosphere and above low-level cloud are also available for studying aerosol global transport. Enhanced remote sensing provides capabilities for the characterization of aerosol type. Major satellite systems that are currently available for studying aerosol intercontinental transport include AVHRR series, ATSR on Envisat, SeaWiFS on Terra, MODIS on Terra, MODIS on Aqua, medium resolution imaging spectrometer on Envisat, atmospheric infrared sounder on Aqua, and ozone monitor instrument on Aqua. It is worth to note that several satellite sensors also provide value-added observations for aerosol global transportation by measuring trace gases such as SO_2, NO_2, and CO (Martin 2008; Streets et al. 2013), which are aerosol precursors or originate from the same sources as particles. These measurements of trace gases can also provide means for identifying source regions of major aerosols and estimating emissions. Qualitative tracking is the essential step of identification of an aerosol plume that follows its passage from one continent to the other. Many satellites can provide visual views for these events in a way that cannot be accomplished by traditional ground measurement systems. After a plume is identified, a progression of quantitative satellite products is required to characterize the aerosol and its transport from images of detected plumes.

8.3 Methods of Estimating the Aerosol Anthropogenic Component

The early effort to quantify the particulate mass from satellite measurement was conducted by using the GOES (Fraser et al. 1984). The mass of an aerosol constituent in a vertical column of the atmosphere is derived from the strong linear regression between the point mass density (ρ) of the fine particle mass and the scattering coefficient of the aerosol (Bs) at low relative humidity (RH)

$$Bs(\lambda) = a + b\rho \tag{8.1}$$

where the scattering coefficient is measured from a certain spectral band λ in the visible spectrum. The regression coefficients a and b depend on the constitute, its region, and the season. The Bs(λ) can be calculated from an aerosol extinction coefficient (Be) from satellite observation following the relationship of

$$Bs(\lambda_1) = g(\lambda)f(rh)\omega_0 Be(\lambda_2) \qquad (8.2)$$

where:
   rh is a relative humidity
   $g \sim \lambda^{-1.6}$ is used to related field measurements of Bs at 480 nm (λ_1) that will be used to the satellite measurement at 610 nm (λ_2)
   ω_0 is the aerosol albedo of single scattering
   f is a RH coefficient, which reduces the scattering coefficient of a hygroscopic aerosol to that of a dry aerosol

The function f is determined by the following restrictions:

$$f = 1.43(1-rh)^{0.7} \quad 0.4 \le rh \le 0.9$$

$$f = 1.0 \qquad\qquad rh < 0.4$$

The equations can be used to estimate the mass density of specific aerosol. For example, the mass density of sulfate ion (SO_4^{-2}) can be obtained by

$$SO_4^{-2} = 0.12f(rh)\omega_0 Be(610) \qquad (8.3)$$

Similarly, the mass density of sulfur (S) in the fine particle mode can be calculated by

$$S = 0.04f(rh)\omega_0 Be(610) \qquad (8.4)$$

Furthermore, the mass of particulate sulfur in a vertical column (M) is obtained by integration of Equation 8.4 with respect to height (z)

$$M(S) = 0.04f \, \omega_0 AOD(610) \qquad (8.5)$$

where:
   AOD is the total vertical optical thickness of aerosol

Although the algorithm could quantify the mass of certain aerosol, large uncertainties exist because of poor accuracy of early satellite measurements. Also, it does not represent well for aerosol anthropogenic component from urban area. Kaufman et al. (2005b) developed a method for a satellite-based estimate of the aerosol anthropogenic component using MODIS data. The method represents the total aerosol optical thickness AOD(550) by its anthropogenic (air pollution and smoke aerosol) (AOD_{anth}), dust (AOD_{dust}), and baseline marine (AOD_{mar}) components:

$$AOD(550) = AOD_{anth} + AOD_{dust} + AOD_{mar} \tag{8.6}$$

The fine aerosol optical thickness, AOD_f, measured by the satellite can be described as follows:

$$AOD_f = f_{550} AOD(550) = f_{anth} AOD_{anth} + f_{dust} AOD_{dust} + f_{mar} AOD_{mar} \tag{8.7}$$

Here, fine fractions of the marine, dust, and anthropogenic component are directly calculated from MODIS data. AOD_{mar} can be determined from surface observation such as AERONET. Furthermore, f_{550} and ADO(550) are derived from MODIS. After rearranging these variables in Equations 8.6 and 8.7, the AOD_{anth} is expressed as

$$AOD_{anth} = \frac{(f_{550} - f_{dust}) AOD(550) - (f_{mar} - f_{dust}) AOD_{mar}}{f_{anth} - f_{dust}} \tag{8.8}$$

The magnitudes of f_{dust}, f_{mar}, and f_{anth} can be determined using the MODIS data that are characterized by the relation of f_{550} and ADO(550). For example, f_{mar} was determined in clean marine region (20°–30°S, 50°–120°E) as $f_{mar} = 0.32 \pm 0.07$ for 2002. Also, f_{dust} value in West of the African coast (15°–20°W, 15°–20°N) was calculated for summer of the year as $f_{dust} = 0.51 \pm 0.03$, and f_{anth} over the western Atlantic (40°–50°N, 70°–90°W) as $f_{anth} = 0.92 \pm 0.03$. To derive f_{anth} and f_{dust}, the marine contribution was subtracted first using Equations 8.6 and 8.7. These values are then used to determine the anthropogenic optical thickness, AOD_{anth}, for each location and time using the MODIS daily measured AOD(550) and f_{550} values. Generally, for the whole of 2002, f_{mar} is determined in clean marine region (20°–30°S, 50°–120°E) as $f_{mar} = 0.32 \pm 0.07$. However, f_{dust} and f_{anth} are determined in West of the African coast (15°–20°W, 15°–20°N) for June–October 2002 as $f_{dust} = 0.51 \pm 0.03$, and over the Western Atlantic (40°–50°N, 70°–90°W for June and 60°–80°W for July) as $f_{anth} = 0.92 \pm 0.03$, respectively. To estimate f_{anth} and f_{dust}, the marine contribution was subtracted first from Equations 8.6 and 8.7. Their values are then used to determine the anthropogenic optical thickness, t_{anth}, for each location and time using the MODIS daily measurements of AOD(550) and f_{550}. Furthermore, the anthropogenic fraction [$A_{frc} = AOD_{anth}/AOD(550)$] derived from Equation 8.8 can be expressed as a function of the aerosol optical thickness. The transformation of the coordinates from [$f_{550}/AOD(550)$] to [$A_{frc}/AOD(550)$] can be used to generate a fixed anthropognenic fraction for the different aerosol types. Moreover, the relation determined by Equation 8.8 can be used to outline the distribution of the anthropogenic aerosol using global data. Aerosols from other nonanthropogenic sources such as dust from Africa and East Asia can be separated from the maps of the total aerosol distribution. This will facilitate the determination of anthropogenic aerosols from urban areas and monitor their global transportations using remote sensing information.

8.4 Applications of Remote Sensing for Global Air Quality Assessment

8.4.1 Characterization of Global $PM_{2.5}$ Distribution

In this example, an annual mean of AOD retrieved from MODIS is used to assess PM air quality over different locations across the global urban areas spread over 26 locations in Sydney, Delhi, Hong Kong, New York City, and Switzerland (Gupta et al. 2006). The high AOD magnitudes observed over Africa are due to dust outflow from Saharan regions. During spring and summer time, high frequency of dust events in dry seasons is mainly responsible for the poor air quality over Africa. In addition to the dust outbreak in the dry season, urban and industrial pollution during summer time in eastern China, northern India, eastern United States, and parts of Europe are also partially responsible for the high concentration of aerosols observed in these regions. Biomass burning in dry season in the southern hemisphere is also visible over South America and Africa.

AOD is calculated first using observed radiances in seven wavelengths (0.47–2.13 µm) over ocean and two wavelengths (0.47 and 0.67 µm) over land and precomputed look up tables provided by Remer et al. (2005). Latitude and longitude information were used to calculate the distance between the MODIS pixel and $PM_{2.5}$ station, so that geographical collocation of MODIS pixels with $PM_{2.5}$ stations is performed. The MODIS level 2 daily AOD products from Terra (MOD04 V003) for the period of January–December 2002 and from Aqua (MYD04 V003) for the period of July–December 2002 is obtained for all study regions. Similar AOD data for Delhi is obtained for the period of July–November 2003. In addition to level 2, MODIS level 3 monthly mean AOD data in $1° \times 1°$ grid resolution from March 2000 to February 2002 is used to assess the spatial distribution of AOD over the globe. Furthermore, all the MODIS pixels within a distance of 0.21 (about 20–25 km) are averaged over $PM_{2.5}$ stations.

The monthly means of $PM_{2.5}$ mass concentration (solid dots) and MODIS AOD (square) for several locations. The thick solid and long dash lines are interpolated lines of $PM_{2.5}$ and AOD, respectively. The 24 h mean values of $PM_{2.5}$ mass (small dotted line) for the entire study period are also shown. The figure includes the air quality categories symbolized by horizontal solid lines along with gray color-coded bar representing air quality condition in each region based on $PM_{2.5}$ mass loading. It is worth to note that these air quality categories are characterized according to the U.S. Environmental Protection Agency (EPA) air quality index; this standard is used for reference purposes and air quality standards in other countries could be different. The air quality category is defined as good if daily mean $PM_{2.5}$ mass concentration is between 0 and 15.4 μg m^{-3}, moderate if it is between 15.4 and 40.4 μg m^{-3}, unhealthy for sensitive groups such as children and older adults if between it is 40.5 and 65.4 μg m^{-3}, unhealthy if it is between 65.5 and 150.4 μg m^{-3}, and very unhealthy if it is between 150.5 and 250.4 μg m^{-3} (United States Environmental Protection Agency 2003).

The MODIS AOD and $PM_{2.5}$ products also represent two different atmospheric loadings of pollutants. The $PM_{2.5}$ is the particle mass concentration with aerodynamic diameter less than 2.5 μm and represents the dry mass of aerosols measured at ground level. The MODIS AOD, however, represents total columnar loading of all aerosol particles from the surface to the top of atmosphere averaged over a large spatial area by 10×10 km^2. It is expected that differences between surface $PM_{2.5}$ and MODIS AOD exist and vary region by region. The relationship between $PM_{2.5}$ mass measured at ground and columnar AOD could not be the same due to variations in source regions and aerosol transportation at different heights. However, some fundamental relationship between these two products is valid in certain domain. For example, vertical profiles of aerosol obtained from observations demonstrate maximum concentrations near-ground and up to the boundary layer (Kaufman et al. 2003), although the spatial distribution may not be in the same pattern in different regions. The analysis also suggested that the mean $PM_{2.5}$ values are less (0–25 μg m^{-3}) over clean environments like Sydney and Switzerland and greater (0–60 μg m^{-3}) over relatively highly polluted regions like Hong Kong, New York City, and Delhi. Occasionally, $PM_{2.5}$ mass values could reach to more than 100 μg m^{-3} over Delhi, and Hong Kong, which could be attributed to possible local aerosol events reported in surrounding areas.

Seasonal variation patterns of $PM_{2.5}$ and MODIS AOD can also be revealed. The seasonal variation in AOD over New York City matches well with the $PM_{2.5}$ trends. Concentration of $PM_{2.5}$ peaked in July with a value of 28 μg m^{-3} of monthly mean, whereas AOD values emerged the highest in June with a value of 0.6, leading to a relatively small the difference between June and July. Furthermore, the air quality in the city remains under the good category throughout the year except during summer months (June–August) when it reaches the moderate category. High values of $PM_{2.5}$ and AOD in July are associated with the transport of smoke from Canada. The monthly mean AOD and $PM_{2.5}$ mass over Switzerland; maximum $PM_{2.5}$ values during winter months and minimum during summer months. A peak AOD value of 0.3 was also observed in April whereas $PM_{2.5}$ peak (35 μg m^{-3}) was seen in January. $PM_{2.5}$ values varied from a high value of 81.3 μg m^{-3} to a low value of 3.3 μg m^{-3} during the period. Therefore, the monthly mean air quality can be categorized in the moderate category throughout the year except in May and July when the air quality is good. The AOD and $PM_{2.5}$ values show similar monthly variations from April through September. Switzerland has many sources of air pollutants such as dust transported from the Sahara Desert but the major part of $PM_{2.5}$ comes from automobile emissions.

The air quality in Hong Kong remains in the moderate category with monthly mean $PM_{2.5}$ values varying from 25 to 40 µg m^{-3} during spring and summer months. The AOD over Hong Kong has a bimodal distribution pattern during the year, whereas $PM_{2.5}$ has a unimodal distribution pattern. The dust transported from the Gobi and Taklamakan Deserts in China elevates AOD concentration in Hong Kong during spring, and impact range of dust transport depends on the prevalence of the northeast monsoon (Duce et al. 1980; Lee and Hills 2003). The maximum hourly $PM_{2.5}$ mass reaches as high as 133.5 µg m^{-3} and the corresponding AOD value becomes 1.1 during the winter season. For urban areas such as Tsuen Wan, Hong Kong, pollution sources are partly local and contributions come from automobile emission and industrial activities.

The monthly mean AOD values are high during summer and low in winter. The peak AOD was observed in May with an AOD value of 0.9 and the minimum was reported as 0.38 in February. The AOD value begins to decline as the summer monsoon approaches Delhi in late July or early August, and the decline in the trend of AOD continues until December when the monsoon rain washout most pollutants. The $PM_{2.5}$ concentration in the Delhi varied from 36 to 52 µg m^{-3} during the time that surface observations were conducted.

The comparison of air qualities in different cities also suggests that the air quality in Sydney remains under the *good* category throughout the year except during in December and January, which is largely contributed to the major bushfire events in the area. Low values of AOD are observed in the southern hemisphere winter and high values in summer seasons. AOD over Sydney exhibits very clear monthly patterns with the maximum of 0.47 in January and decreases till June to 0.1. After then, it increases up to December. The annual mean of $PM_{2.5}$ mass ranges from 11.0 ± 7.0 to 15.0 ± 5.8 µg m^{-3} over different stations in the region, which is within the U.S. EPA standards. The maximum hourly $PM_{2.5}$ values ranging from 50.0 to 76.8 µg m^{-3} were detected during the January 2002 bushfire event in the area.

The relationship between hourly $PM_{2.5}$ and MODIS AOD was analyzed by calculating the linear correlation coefficients. The 24-h averaged $PM_{2.5}$ and AOD from Terra and Aqua MODIS products over five different metropolitan areas around the world was also revealed in the five study regions. These 24-h averaged $PM_{2.5}$ concentration measurements are used to define air quality indices at each local station. The simple regression equation between MODIS AOD and $PM_{2.5}$ mass concentration is developed by dividing the 24-h mean $PM_{2.5}$ into 11 bins of 5 µg m^{-3} intervals. Therefore, the air quality categories can be connected with satellite data. The figure also shows these bins with averaged points in MODIS AOD and $PM_{2.5}$ space as black dots and the solid black line shows the regression line between these two parameters. The height of black box is for the standard deviation in MODIS AOD for a particular $PM_{2.5}$ bin. The correlation between bin-averaged AOD and $PM_{2.5}$ concentration is very high. The empirical relationship between AOD and $PM_{2.5}$ mass can be used to quantified $PM_{2.5}$ concentration and an estimate of the air quality. Although there are possible uncertainties in $PM_{2.5}$ measurements (usually in the order of 0%–10%) due to sampling methods and instrument-related issues, as well as uncertainties in MODIS AOD retrievals, linear regression is still a useful tool to estimate local air quality conditions. In the global urban area analysis, the highest linear correlation coefficient (0.85) was found over one station in Delhi, and it was lowest (0.11) at the Richmond station in Sydney due to the very low mean optical thickness. Moderate to high correlation coefficient (0.6) is found over New York City region (Gupta et al. 2006). The general linear regression equation for MODIS AOD(550) and $PM_{2.5}$ can be expressed as

$$AOD(550) = 0.006 \times PM_{2.5}\left(\mu g\ m^{-3}\right) + 0.149 \qquad (8.9)$$

Equation 8.9 has a linear correlation coefficient about 0.96 for both Terra and Aqua satellites. $PM_{2.5}$ concentration can be quantified using MODIS AOD from this regression relation, and further for an estimate of the air quality index. The use of linear equation (Equation 8.9) to estimate $PM_{2.5}$ concentrations in different regions results in different linear correlation coefficients between hourly $PM_{2.5}$ and MODIS AOD, which are as follows: 0.6 for New York, 0.14 for Switzerland, 0.40 for Hong Kong, 0.41 for Delhi, and 0.35 for Sydney. The satellite–$PM_{2.5}$ correlations could vary from region to region.

8.4.2 Intercontinental Transport of Pollutant

Satellite measurements, besides being used to estimate columnar atmospheric density, have been used to estimate the intercontinental transports of dust mass, combustion aerosol mass (including industrial pollution and biomass burning smoke), and pollutant mass. Satellite measurements for assessing aerosol transports can provide source information such as where aerosol sources are from and potential impact extents of pollutants. Generally, the estimates of aerosol fluxes in continental outflows and inflows are accomplished by a three-step approach (Yu et al. 2013a). In the first step, satellite measurements of AOD and aerosol microphysical properties (e.g., size and shape) are used to distinguish mineral dust from combustion-related aerosol. In the second step, AOD for dust or combustion aerosol is converted to aerosol (dry) mass concentration profile by characterizing aerosol hygroscopic property from *in situ* measurements and aerosol vertical distributions from LIDAR measurements (Yu et al. 2008, 2010, 2013b). In this step, pollution aerosol optical depth, AODp, is converted to dry mass loading Mp as

$$Mp(1) = \frac{AODp(1)}{f[RH(1)]MEEp}$$ (8.10)

where:

MEEp $(m^2\ g^{-3})$ is dry mass extinction efficiency of pollution aerosol

f(RH) is a function that includes an increase of aerosol extinction with increasing RH

In the third step, the aerosol mass concentration in combination is used to estimate the zonal export and import of aerosol with zonal wind speed. Seasonal, annual, and latitude-integrated flux variations of pollution fluxes can be estimated using the model with MODIS/Terra data and climate records.

The MODIS-estimated seasonal pollution fluxes across continents in 2004 indicated that both eastern Asia outflow and North America inflow peaked in spring, with the flux of 6.8 and 1.7 Tg/season, respectively. The seasonal variations suggest that the outflow and inflow are weakest but not negligible in summer, with a magnitude of about 1/3 of the springtime maximum and 10% of the annual flux. Moreover, winter and autumn appear to be transitional seasons when pollution fluxes vary between the springtime maximum and the summertime minimum. Over the western Pacific, the wintertime pollution flux is about 30% higher than that in autumn. The analysis of interannual variations of pollution fluxes shows that the pollution largely from China influenced the West Pacific region to the greatest extent in March, April, and May (MAM) and June, July, and August (JJA) in 2005. The AOD in the high-latitude North Pacific (40°–60°N) was increased to a level that is more than a factor of 2 larger than in other years in MAM and JJA in 2003. Both the total fire pixels and burned area are a factor of 3–4 larger than that in 2004, resulting in that the burned area in 2003 was the largest in a decade. On the other hand, boreal forest fires in the North American continent (Alaska and Canada) in summer were more intense in 2004 than in 2003. The MODIS pollution aerosol map shows that these boreal fires, which mainly influence the North American continent in the eastern and southeastern United States, also produce a discernible influence on the surrounding northeastern Pacific. In the mid-latitude West Pacific, pollution largely originated from China influenced the region to the greatest extent in MAM and JJA in 2005. In the tropical East Pacific, however, the springtime biomass burning smoke from Central America was observed as the weakest in 2004.

To illustrate intercontinental distribution of aerosol, a global view of aerosol depth and size is displayed in Figures 8.2 and 8.3. These graphics show average monthly aerosol amounts including AOD and percent of small size aerosol around the world based on observations from the MODIS on NASA's Terra satellite (http://earthobservatory.nasa.gov/GlobalMaps). The unit of AOD varies from less than 0.1, indicating a crystal clear sky with maximum

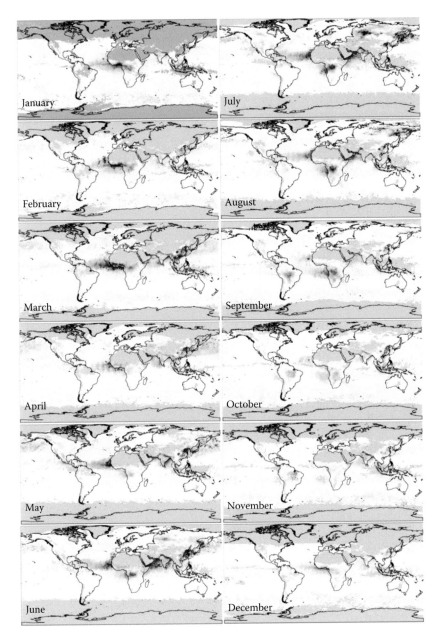

FIGURE 8.2
Global distributions of monthly mean of aerosol optical depth developed from NASA Earth Observatory. The color legend is the same as in Figure 8.1a. (Data from NASA Earth Observatory, http://earthobservatory.nasa.gov/GlobalMaps.)

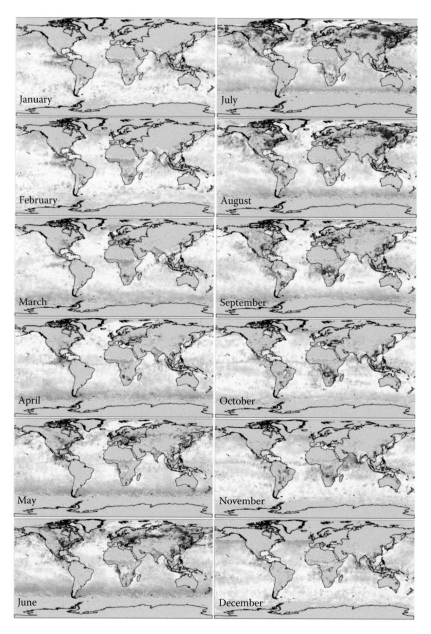

FIGURE 8.3
Global distributions of monthly mean of aerosol size developed from NASA Earth Observatory. The color legend is the same as in Figure 8.1b. (Data from NASA Earth Observatory, http://earthobservatory.nasa.gov/GlobalMaps.)

visibility, to a value of 1, indicating very hazy conditions. The percentage of small particles represents the proportion of anthropogenic-induced pollutant.

High aerosol amounts usually occurred over South America from July to September (Figure 8.2). This pattern is associated with land clearing and agricultural fires that are widespread across the Amazon Basin and Cerrado regions during the dry season. Aerosols have a similar seasonal pattern in Central America (March–May), central and southern Africa (June–September), and Southeast Asia (January–April). However, some high aerosol concentrations are related to other events, such as around the Arabian Peninsula and nearby oceans due to dust storms during summer time. Air pollutants produced from anthropogenic activities elevates aerosol amounts over eastern China, in northern India, and in eastern United States.

Aerosol particles of natural origin usually have a larger radius than aerosols generated by anthropogenic activities. The false-color maps, which are generated from MODIS on Terra satellite, illustrate locations where natural aerosols, anthropogenic pollutants, or a mixture of both on a monthly basis (Figure 8.3). Green areas represent aerosol plumes dominated by larger particles, while red areas characterize aerosol plumes dominated by small particles. Yellow areas are for plumes of mixing large and small aerosol particles and gray is for areas where the sensor did not collect data. Over land, several places are dominated by smaller aerosols. Areas where human-triggered or natural fire activities are common from August to October can be seen in northern Canada. However, anthropogenic-induced pollution is largely accountable for small aerosols in areas including the eastern United States, Europe, and eastern Asia in summer.

8.5 Summary

Human-dominated landscapes have unique biophysical characteristics. Humans redistribute organisms and the fluxes of energy and materials. The use of satellite-based measurements, especially in EOS-era, provides information of trans-Atlantic transport of dust mass, trans-Pacific transport of combustion aerosol mass (including industrial pollution and biomass burning smoke) and dust mass, and cross-Mediterranean Sea transport of European pollution mass. The satellite-based estimates of aerosol import and export in the zonal direction have provided important insights into aerosol intercontinental transport and its impacts on regional air quality, climate change, and biogeochemical cycle. However, it is necessary to point out that the satellite-estimated mass fluxes are subject to large uncertainties. Major factors that introduce uncertainties include satellite-derived component AOD, mass extinction efficiency, hygroscopic properties, transport height, and satellite sampling. The uncertainties can be reduced through a synergy of multi-sensor measurements from on-orbit satellites (e.g., A-Train synergy). Aerosol type characterization using information of particle size and shape can be cross-examined with trace gas measurements and traced to source locations through the backward trajectory analysis. Further research is expected to integrate satellite measurements and aerosol transport models for a comprehensive characterization of aerosol intercontinental transport. Long-term trends of the transcontinental pollution transport and its environmental impacts could be detected and characterized with an accumulation of satellite data in the future.

9

Urban Heat Island and Regional Climatic Effect

9.1 Introduction

Climate represents the average pattern of variation in temperature, precipitation, atmospheric pressure, wind, and other meteorological variables in a given region over long periods of time. The climate of a region generated by the climate system has five components: atmosphere, hydrosphere, cryosphere, land surface, and biosphere. The change of land surface can modify local or regional climate by alternating transition of energy fluxes from surface to atmosphere. The transition is completed through both air dynamic and thermal processes. The terrain feature of urban surface can change airflow and creates interacting wakes and plumes (of heat, humidity, and pollutants) introduced by individual roughness elements (Arnfield 2003). The urban atmosphere usually is divided into two layers: mixing layer and surface layer. The mixing layer is from the top of urban construction to the top of atmospheric boundary. The surface layer is from land surface to the top of urban construction. Urban structures and features such as buildings, streets, and parks significantly shape weather patterns by altering atmospheric flows, turbulence, humidity, precipitation, and temperature, and influence the transport, dispersion, and deposition of atmospheric pollutants (Cotton and Pielke 1995; Taha et al. 1997). Beyond the air dynamic effects introduced by urban landscape, urban areas also implement surface radiative and thermodynamic impacts on atmospheric system (Voogt and Oke 2003; Xian 2008). Urban development changes natural landscape feature into anthropogenic impervious surface that usually is a poor storage system for water and is a heat source. The resultant surface temperature in urban area affects thermal admittance and surface emissivity, the radiative input at the surface from the sun and atmosphere, and the near surface atmosphere condition and turbulent transfer from the surface.

The patchy and heterogeneous nature of the urban surface has significant implications in the interpretation of diverse energy budgets that generate contrasts in surface characteristics and lead to mutual interactions by radiative exchange and small-scale advection. One of the effects associated with urban land surface is urban heat island (UHI), which is a result of the evaluation of air temperature near surface within cities and built-up areas relative to the surrounding rural areas. UHI effect is related to the retention of solar heat in the buildings, ground surfaces, and the obstruction and re-absorption of nighttime outgoing long-wave radiation by surrounding structures. Paved land surfaces absorb more solar radiation downward than soil and vegetation surfaces. The heat is then available for release overnight. Reduced ventilation can hinder the dispersal of urban heat and enhance the heat island effect. The physical properties of buildings and other structures, and the emission of heat by human activities also intensify UHI. Generally, UHI are most pronounced on clear, calm nights; their strength depends also on the background geography and climate, and there are often cool islands in parks and less developed areas. Heat islands can affect communities by increasing summertime peak energy demand, air conditioning costs, air pollution and greenhouse gas emissions, heat-related illness, and mortality.

The characteristics of the UHI effect have been studied extensively in different approaches (Arnfield 2003). Traditionally, the UHI intensity for a given urban area is studied by comparing the air temperature difference between a given urban site or several sites and a carefully selected nearby nonurban site and multiple nonurban sites from meteorological observations (Oke 1973; Landsberg 1981). However, the representativeness of observational sites is quite important for accurately estimating UHI intensity that usually is influenced by many factors, such as land cover type, urban extent, urban population, climate background, latitude, height above sea level, and topography. The construction features, such as built-up and nonbuilt-up ratios, green surface ratio, and sky view factor, can also affect spatial distribution and intensity of UHI. Therefore, the selection of representative observation sites is a challenge. Additionally, the nonurban sites might be merged into the urban area in a certain time period; such change could influence continuous UHI examination.

Remote sensing technology, especially thermal remote sensing, has been used to detect urban thermal characteristics since the early 1980s (Carlson et al. 1981; Roth et al. 1989). Many studies focused on remotely sensed land surface temperature (LST) in urban areas (Weng et al. 2004; Xian and Crane 2006). Thermal remote sensing of urban surface temperatures observes urban LST variation in response to the surface energy balance. The resultant surface temperature incorporates the effects of surface radiative and thermodynamic properties, including surface moisture, thermal admittance and surface emissivity, the energy exchange between surface and atmosphere, and the effects of the near-surface atmosphere condition. Subsequently, many studies of UHI based on remote sensing technology have been accomplished using LST as an important parameter (Tomlinson et al. 2011). Most studies found that LST modulates the air temperature of the lower layer of the urban atmosphere and is a primary factor in exploring surface radiation and energy exchange, the internal climate of buildings, the spatial structure of urban thermal patterns and their relation to urban surface characteristics, surface-air temperature relationships, and human comfort in cities.

In this chapter, the methods used to derive LST from thermal remote sensing are introduced in Section 9.2. The application of remote sensing-derived LST and its influence on surface air temperature is illustrated in Section 9.3. The global view of UHI effect is introduced in Section 9.4. The regional climate implication of UHI in urban areas is discussed in Section 9.5. Section 9.6 is the summary of the chapter.

9.2 Calculation of LST from Remote Sensing

A fundamental function of sensors on a remote sensing platform is the detection of electromagnetic radiation. This is useful as different objects emit radiation in different ways so that the spectral response can be analyzed. Within the solar radiation spectrum, the most useful for LST measurements is the thermal infrared (TIR), between 8 and 15 μm. There are a number of different satellite remote sensing platforms with multiple sensors in the TIR spectrum. Datasets are available for different time periods at different resolutions and with varying accuracy.

9.2.1 AVHRR Instrument

The advanced very-high-resolution radiometer (AVHRR) sensor, having been boarded on a number of National Oceanic and Atmospheric Administration (NOAA) satellites and currently operational on NOAA 15–19, provided daily daytime coverage. The spatial resolution is about 1.1 km and LST is derived from TIR channels 4 (10.3–11.3 μm) and 5 (11.5–12.5 μm), with a global dataset provided through the sun-synchronous orbit.

Data from the AVHRR sensor has a relatively long historical record and global coverage. Therefore, AVHRR data has been used by many different users for different applications. Such long-term record is not possible with most other sensors as historical data are not available.

LST can be derived from the AVHRR instrument using TIR channels. First, surface radiance values are calculated with the use of calibration coefficients contained within the metadata data. These radiance values are then corrected for the nonlinearity of AVHRR channels 4 and 5 with the use of radiance coefficients. The corrected radiance values (R_i) are then converted to brightness temperature using Planck's equation of radiation:

$$T(R_i) = \frac{C_1 v_i}{\ln\left[1 + \left(C_2 v_i^3 / R_i\right)\right]} \tag{9.1}$$

where:

$C_1 = 1.438833 \times 10^{-5}$ cm K

$C_2 = 1.1910659 \times 10^{-5}$ mW m^{-2} sr^{-1} cm^4

v_i is the central number of each channel

The surface temperature is calculated from the brightness temperature using the following relation (Price 1984):

$$T_{sfc} = \frac{T_4 + 3.339(T_4 - T_5)(5.5 - \varepsilon_4)}{4.5 + 0.75T_4(\varepsilon_4 - \varepsilon_5)} \tag{9.2}$$

where:

T_4 and T_5 are brightness temperatures in channels 4 and 5

ε_4 and ε_5 are emissivity coefficient for channels 4 and 5

By assuming a surface emissivity of 1.0, Equation 9.2 is simplified to

$$T_{sfc} = T_4 + 3.3(T_4 - T_5) \tag{9.3}$$

Alternatively, it is possible to fit the heat island to the Gaussian surface (Streutker 2002; Streutker 2003). This is accomplished by performing a least-squares fit to the natural logarithm of the temperature. The fitting surface was chosen to be a two-dimensional Gaussian superimposed on a planar rural background. The form of this surface defined as

$$T(x,y) = T_0 + a_1 x + a_2 y + a_0 \exp\left[-\frac{(x - x_0)^2}{2a_x^2} - \frac{(y - y_0)^2}{2a_y^2} \right] \tag{9.4}$$

The method provides a measure of the UHI magnitude for the entire city (a_0), the spatial extent (a_x and a_y), and central location (x_0 and y_0) of the heat island as well. This method is also used to determine the spatial gradients (a_1 and a_2) and mean value (T_0) of the surface temperature of the surrounding rural area.

It was necessary to mask the water and cloud areas in order to isolate the heat island in the image. Furthermore, the areas in the general urban areas were masked in order to quantify the constant and linear components. A least-squares planar fit to the remaining rural data was performed to determine the constant and linear components of temperature. Once the constant and linear temperature components were subtracted, the heat island can be fitted to the Gaussian surface by performing a least-squares fit to the natural logarithm of the temperature to

$$T(x,y) = a_0 e^{\left[-a_x(x - x_0)^2 - a_y(y - y_0)^2 \right]}$$

A correlation coefficient was calculated using the measured surface temperature and Gaussian surface.

Furthermore, it was found that heat island magnitude is negatively correlated with rural temperature. The functional relationship can be represented by Equation 9.5 as follows:

$$UHI = 3.56°C—0.044T_{rural} \tag{9.5}$$

This relation reveals that the rural temperature can also be used to quantify the intensity of UHI.

9.2.2 Landsat Imagery

The Landsat series of satellites launched in the early 1980s has the longest record of Earth observations from space. The Thematic Mapper (TM) on Landsat 5 had a visible resolution of 30 m and a TIR resolution of 120 m (band 6, 10.4–12.5 µm). Landsat 5 stopped data collection in 2011. Landsat 7's Enhanced TM (ETM+) launched in 1999 collects thermal data at a 60 m resolution (also with band 6, 10.4–12.5 µm). Both Landsat 5 and 7 have a near polar sun synchronous orbit with a revisit time of 16 days, indicating that at a given point, Earth should be imaged at approximately the same local time (~1000 h) every 16 days. The ETM+ offers some of the highest-resolution in thermal measurements from space, and data is available freely from the U.S. Geological Survey (http://earthexplorer.usgs.gov/).

Usually, radiance values from TM band 6 and ETM+ band 6 H, saturating at 322 K, can be transformed to radiant surface temperature values for Tampa Bay. However, in areas where landscapes are dominated by bare and sandy soil, ETM+ band 6L, saturating at 347.5 K, is used. The thermal band was first converted from DN to at-satellite radiance using Equation 9.6 and then to effective at-satellite temperature (T_b), which is also called *brightness temperature* or *radiant surface temperature*.

Thermal band imagery can also be converted from spectral radiance (as described above) to a more physically useful variable. This is the effective at-satellite temperature of the viewed Earth-atmosphere system under an assumption of unity emissivity and using prelaunch calibration constants. The conversion formula is used to calculate the effective at-satellite temperature or radiant surface temperature from thermal band:

$$T_b = \frac{K2}{\ln\left[(K1/L_\lambda)+1\right]} \tag{9.6}$$

where:
$K2$ is the calibration constant 2 (1260.56 for TM and 1282.71 for ETM+)
$K1$ is the calibration constant 1 (607.76 for TM and 666.09 for ETM+)
L_λ is defined in Equation 3.6
T_b is effective at-satellite temperature in Kelvin

The radiant surface temperature can be used to estimate LST and therefore has been widely used to evaluate UHI effect around the world.

9.2.3 MODIS

The moderate resolution imaging spectroradiometer (MODIS) sensor carried on both NASA's Aqua and Terra satellites that have near-polar orbits resulting in two images per satellite per day. Images from the MODIS have a spatial resolution of 1 km for its useful LST products. These LST products primarily use TIR bands 31 (10.78–11.28 μm) and 32 (11.77–12.27 μm) combined with split-window algorithms (Wan and Dozier 1996) that have been tested by multiple studies with results suggesting accuracies greater than 1 K over homogeneous surfaces (Wan 2002, 2008; Wan et al. 2004; Coll et al. 2005). The method has several major advantages. It does not need profiles of atmospheric temperature and water content. The atmospheric effects are corrected based on the differential absorption in adjacent TIR bands instead of absolute atmospheric transmission in a single band. Therefore, the method is less sensitive to the uncertainties in optical properties of the atmosphere.

A typical linear split-window algorithm can be written as

$$\text{LST} = a_0 + a_1\left(T_i + T_j\right) + a_2\left(T_i - T_j\right) \tag{9.7}$$

where:
$a_k(k = 0,1,2)$ are coefficients that depend on the spectral response function of the thermal channels i (channel 31) and j (channel 32), the two channel emissivities

Several MODIS LST products have been produced using information of onboard sensors Terra (MOD) and Aqua (MYD). These products include LST/emissivity daily in 1 km (MOD11A1 and MYD11A1), LST/emissivity eight-day average in 1 km (MOD11A2 and MYD11A2), LST/emissivity daily in 6 km (MOD11B1 and MTD11B1), LST/emissivity daily at 5-minute increments swath-based in 1 km (MOD11_L2 and vMYD11_L2), LST/emissivity daily in 0.05° (MOD11C1 and MYD11C1), LST/emissivity eight-day in 0.05° (MOD11C2 and MYD11C2), and LST/emissivity monthly in 0.05° (MOD11C3 and MYD11C3). These products have been used to quantify LST in urban areas in both regional and global scales.

9.3 Quantification of UHI Using Remotely Sensed LST Products

One of intensive uses of remotely sensed LST products is for urban area. The urban environment with higher temperature impacts human health, ecosystem function, local weather, and potentially climate. Generally, the

UHI phenomenon is caused by a reduction in latent heat flux and an increase in sensible heat in urban areas as vegetated and evaporating soil surfaces are replaced by relatively impervious low albedo paving and building materials. Such surface land cover change usually results in a positive temperature difference between urban and surrounding nonurban areas. The UHI effect can be quantified using both air and surface temperatures. Generally, UHI defined by the air temperature has a strong diurnal cycle and is more important at night. Surface temperatures can also characterize the urban heat phenomenon. However, surface temperatures can be both higher and more variable than concurrent air temperatures due to the lower albedo and complexity of the land cover types in urban environments and variations in urban topography. These features do not preclude them to be used to analyze the UHI effect. Instead, surface temperatures are more easily related to surface conditions and the greatest surface temperatures are observed during midday versus nighttime for air temperature due to quickly surfaces heating or cooling.

9.3.1 UHI Effects in a Local Scale

Remotely sensed data of LST, vegetation index, and other surface characteristics have been widely used to describe the UHI phenomenon. Satellite remote sensing provides a straightforward and consistent way to determine thermal difference between urban and rural area. The use of remotely sensed data for UHI study can provide details of both spatial distribution and intensity of LST across urban and nonurban areas. One approach for assessing UHI effect is to integrate land cover and LST to quantify both urban extent and thermal difference between urban and rural (Xian and Crane 2006; Xian 2008). The method has been implemented for two metropolitan areas: Las Vegas and Tampa Bay in the United States. Urban extent was quantified from percent impervious surface that was derived from Landsat data using the method described in Chapter 3. LST was directly calculated from radiant temperature using Landsat TM thermal band. To isolate land thermal features in each image, water was masked out prior to LST estimation. After the water pixels were removed, temperatures for the remaining pixels were re-mapped according to each impervious surface area (ISA) group. Temperature histograms were created for each LST map. Some extremely high and low temperatures associated with ISA had been retained, for example, low temperatures caused by shadows in high-density urban areas and high temperatures caused by image noise. The final radiant surface temperature map was produced by removing pixels that have temperatures greater or less than two standard deviations (2σ) from the mean.

The mean radiant temperature ($\overline{T_b}$) for imperviousness in every 10% was calculated and further aggregated to three ISA categories—10%–40%, 41%–60%, and >60%, which are considered urban areas, to calculate differences from $\overline{T_b}$ for ISA percentages less than 10, which are counted as rural areas.

Figure 9.1 shows the percent ISA associated mean radiant temperature in the Tampa Bay watershed (Figure 9.1a) and Las Vegas Valley (Figure 9.1b). Apparent differences between for ISA ≥ 10% and for ISA < 10% exist in the two areas. In the Tampa Bay area, difference of $\overline{T_b}$ between urban and rural areas is about 1.9°C. In the Las Vegas area, the difference of $\overline{T_b}$ between urban and rural areas is about –0.4°C. The core urban area is cooler than the surrounding rural areas where gravel and the bare sandy soils are dominated land cover types and can be heated quickly in daytime. In addition to the difference in mean radiant temperature between urban and rural area, variations in the magnitude of $\overline{T_b}$ in different urban development density areas compared to rural areas were also quantified. In the Tampa Bay, the difference between the $\overline{T_b}$ for high-density urban areas when compared to the $\overline{T_b}$ for rural areas reached 2.41°C. The standard deviation values of $\overline{T_b}$ were large for areas with percent ISA greater than 60, suggesting that these areas experience wide variation in $\overline{T_b}$ because of different building structures and construction materials. The standard deviation values of $\overline{T_b}$ were relatively small for areas with ISA between 10% and 40%. The homogeneity of construction types in residential areas contributed to this relatively low variation in $\overline{T_b}$. In Las Vegas, the thermal gradient from the center to the edge of Las Vegas is therefore summarized as transitioning from hot (high percent ISA) to warm (medium percent ISA) to cool (low percent ISA). The $\overline{T_b}$ for the nonurban area is higher than that for the urban area, but lower than that for the high-density urban area. In general, urban land use and land cover (LULC) creates a daytime cooling effect in low- to medium density urban areas and UHI in high-density urban areas in Las Vegas.

The influence of urban vegetation on UHI can be determined by comparing NDVI and temperature difference between urban and rural areas in Las Vegas and Tampa Bay. The mean normalized difference vegetation index (NDVI) for different ISA categories is estimated for these two areas. Mean NDVI and the associated standard deviation for different percent ISAs were also calculated for the Tampa Bay watershed in 1995 and 2002. Generally, smaller NDVI values were obtained for urban areas than for nonurban areas, and there is a consistent decrease in the mean NDVI as the percent imperviousness increases. Lower imperviousness has higher NDVI values because of the predominance of vegetation cover in these areas. In contrast, high-density urban areas, including commercial, industrial, and some residential developments, are composed of less vegetation and lower NDVI values. Usually, the standard deviation of NDVI in the urban area is higher than that for the rural area because vegetation canopy associated with landscaping has greater variation than natural vegetation, resulting in a sparser pattern in urban vegetation.

This phenomenon is more apparent in the high-density ISA category. For example, in the Tampa Bay watershed, the dense natural vegetation canopy reduces surface radiant temperature measured by the sensor and causes a relatively low temperature and higher NDVI in rural areas. As ISA increases from low- to high-urban development density, mean NDVI decreases, and

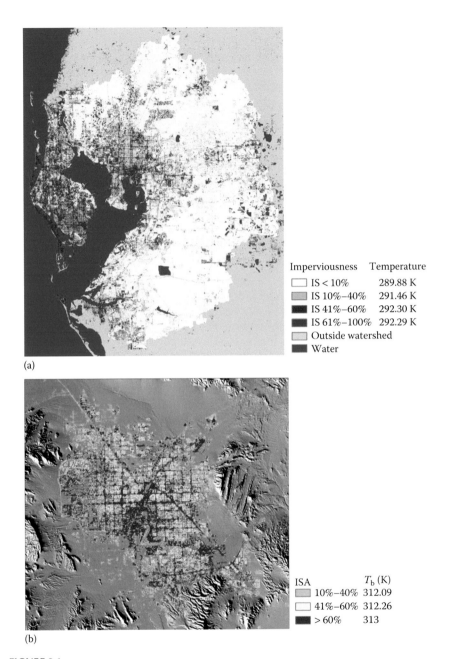

(a)

Imperviousness	Temperature
☐ IS < 10%	289.88 K
▨ IS 10%–40%	291.46 K
■ IS 41%–60%	292.30 K
▦ IS 61%–100%	292.29 K
☐ Outside watershed	
■ Water	

(b)

ISA	T_b (K)
▨ 10%–40%	312.09
☐ 41%–60%	312.26
■ > 60%	313

FIGURE 9.1
Surface temperatures characterized by impervious surface in Tampa Bay (a) and Las Vegas (b) areas.

temperature gradually increases. This results in a negative, moderately low correlation between mean radiant temperature and NDVI in the watershed.

In contrast, the environs of Las Vegas and land cover types are very different from the Tampa Bay watershed. Natural vegetative cover in rural areas is sparse. In urban areas, most vegetation canopies in residential areas including trees, bushes, and grasses are planned by people. Vegetation found in business and commercial areas is composed of lawns, shrubs, and a few trees with little natural canopy. Mean NDVI for different ISA categories in Las Vegas had a reverse distribution compared to that in the Tampa Bay watershed. Urban areas have more vegetation canopy coverage than rural areas.

Las Vegas urban areas also have larger standard deviations than that in rural areas, indicating that more vegetation patchiness exists in urban areas. Vegetation density was the greatest for areas characterized by a medium density ISA of 41% to 60%. Vegetation coverage in downtown, business and commercial centers was lower than that in low-density urban areas. Areas with a low-density ISA have the greatest vegetation variation, suggesting that the canopy is composed of a greater variety of species. Therefore, a cooling effect is found in the low- and medium-density ISA areas. Such cooling effect in these areas appears to be caused by vegetation canopy cover, especially by tree canopies. High average NDVI brought down temperatures by half to one degree in these urban areas. In the high-density ISA areas, especially in downtown areas, the predominance of large parking lots, tall hotel buildings, and entertainment gaming facilities is believed to be a major factor in the rise of mean radiant temperature in these areas. Vegetation canopies from landscaping in these areas are unable to counter to balance the high-density ISA areas to be cooler than the bare and rocky soil surroundings.

9.3.2 Analysis of UHI Using Gridded Temperature and Land Cover Datasets

The use of impervious surface data to analyze UHI effect was further performed using gridded temperature to examine the temperature differences between the urban and rural locations for several levels of urbanization as defined by the level of ISA in Dallas, Texas (Gallo and Xian 2014).

The mean daily temperatures were computed from the daily minimum and maximum temperatures available from the Daymet (Thornton et al 1997) daily surface gridded dataset (available from Thornton et al. 2012) and were used in the analysis with the spatial resolution of 1×1 km^2. Daymet uses a series of algorithms to interpolate and extrapolate station observations of daily maximum and minimum temperatures to produce daily gridded estimates of temperature (http:/dayment.ornl.gov/overview). The daily mean air temperatures were averaged over each month for the 2006–2011 interval.

The U.S. Geological Survey National Land Cover Database (NLCD) 2006 impervious surface product was used to define different urban land cover levels. The original 30 m ISA was degraded to 1 km to match the spatial resolution of the gridded climate data. The ISA was used to define each 1×1 km^2 grid cell as either urban or rural based on the level of ISA within each grid cell. Three ISA levels were used in this study and included ISA levels of $\geq 5\%$ (ISA5), 25% (ISA25), and 50% ISA (ISA50). Subsequently, these ISA levels were used as thresholds to define the 1×1 km^2 grid cells as urban (grid cell ISA values that met or exceeded the listed thresholds) or rural (values less than the listed thresholds). Water was excluded from rural areas through use of data provided in the NLCD 2006 land cover product. Figure 9.2 shows

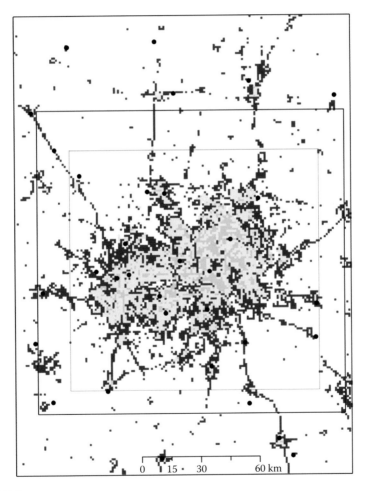

FIGURE 9.2
The Dallas, Texas, metropolitan area with three regions identified (region 1 is the outermost rectangle) that were included in spatial analysis of urban and rural temperatures. Green, yellow, and red designate the ISA5, ISA25, and ISA50, respectively, levels of urbanization. Black (closed) circles are locations of GHCN-daily meteorological stations around the area.

the Dallas metropolitan area and three regions extending from core urban areas. Different urban intensities of ISA5, ISA25, and ISA50 are outlined in different colors. The locations of the Global Historical Climatology Network-Daily Database (GHCN; Menne et al. 2012) stations around the area are also displayed. Urban extents defined by the three ISA thresholds were between 7365 and 710 km^2 in region 1, between 6631 and 706 km^2 in region 2, and between 6083 and 701 km^2 in region 3.

For different urban categories, as the ISA level increased, the urban and rural temperatures increased. These temperature patterns agreed with the results of previous research (e.g., Arnfield 2003) that documents the observation of generally greater temperatures at urban locations compared to rural locations. The observed higher urban temperatures appear to be associated with the greater concentration of highly impervious grids within the ISA50 urban cells. Also, the ISA50 threshold likely resulted in numerous *near-urban* or suburban grid cells (up to 49% impervious) being classified as rural, which led to the relatively high (compared to other ISA levels) rural temperatures. Alternatively, at the ISA5 level many *nearly rural* grid cells were likely included with those defined as urban, while those grid cells classified as rural were likely to be truly rural grid cells. This classification resulted in relatively low temperatures observed for both the urban and rural classified stations. Furthermore, the size of the region included in the analysis also influenced the observed urban and rural temperatures. Region 1, for example, included the largest spatial area (the largest number of grid cells) and resulted in the greatest differences in urban and rural temperatures. Region 3 included the least spatial area that was highly urbanized (40% urban at the ISA5 level) revealed the least difference in the urban and rural temperatures.

Additionally, urban and rural average monthly temperatures observed for each region and their differences were analyzed for grid cells defined as urban based on an ISA level greater than 5% (Figure 9.3). Generally, the urban-designated grid-based temperatures were greater than the rural temperatures and resulted in positive differences between the urban and rural temperatures for all months. It is not surprising to find that the greatest urban–rural differences in temperatures were observed in the analysis of region 1. The mean monthly urban–rural temperature differences over the 2006–2011 interval for the three regions were significant (p value $= 0.01$) and varied from 0.86°C (region 1), to 0.48°C (region 2), and 0.34°C (region 3). Figure 9.3 also illustrates that urban–rural temperature differences in region 1 reached the maximum between early summer and fall months between June and September. The minimum differences usually appeared in winter months between January and March. The monthly temperature differences in regions 2 and 3 follow similar pattern but slightly different in magnitudes from those in region 1. The maximum differences in both regions 2 and 3 appeared in winter months between January and March, except May in 2010. The minimum differences, however, appeared

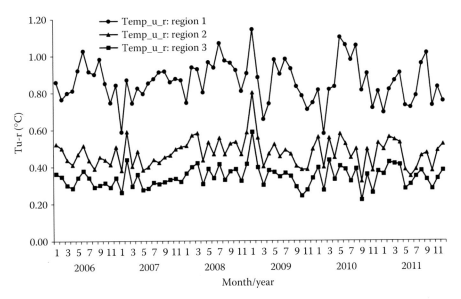

FIGURE 9.3
Monthly urban–rural temperature differences calculated from Daymet grid cell temperatures between 2006 and 2011 for the three regions of the Dallas area. The ISA5 threshold is used to define urban areas.

TABLE 9.1

Annual Urban–Rural Temperature Differences Averaged Over the 2006–2011 Interval for the Three Subregion with Three Thresholds for Grid Cell Classification of Urban or Rural

	$T°C$ (Urban–Rural)		
Region	ISA5	ISA25	ISA50
1	0.86	0.96	0.96
2	0.48	0.52	0.50
3	0.34	0.35	0.32

in August 2006, January 2007, April 2008, November 2009, September 2010, and June 2011. Such temporal variations for urban and rural temperature differences remain similar for all three ISA levels, although the magnitudes differed slightly. The annual urban and rural temperature differences were summarized in Table 9.1, suggesting higher differences in urban areas than in rural area patterns.

Similar to the change in the number of grid cells (and associated area) designated urban or rural in the grid-based analysis, the number of climate stations were designated as urban or rural changed based on the same ISA thresholds.

There were 14 urban and 19 rural stations at the ISA5 threshold level in the area. The number of urban-designated stations decreased as the ISA threshold increased. Only nine stations were designated urban (24 rural) at the ISA25 level and only one station was designated urban (32 rural) at the ISA50 level.

The GHCN station and Daymet data generally agreed in that the urban stations (or associated grid cells) had greater temperatures than the temperatures observed at the rural stations. However, there was greater variation in the urban–rural differences observed in the station data compared to the Daymet data, especially a greater range in the monthly differences was observed between the urban and rural station data compared to the Daymet values. The results of this analysis of urban and rural mean temperatures suggest that similar analyses are warranted for gridded minimum and maximum temperatures in conjunction with gridded impervious surface area data.

9.3.3 UHI Effect across the Continental United States

The spatial character of UHI effect across the continental United States has been assessed using a combination of satellite and ecological map data (Imhoff et al. 2010). The relationships between percent ISA and LST across many cities are analyzed by calculating seasonal UHI for cities in similar ecological settings, and comparing the amplitude of the UHI for the major biomes. To compare urban places within and between settings, the terrestrial ecoregions map developed by Olson et al. (2001) was used to stratify the analyses and constrain the sampling around each urban area according to its biome. Totally, eight biomes were divided, including temperate broadleaf and mixed forest (northern group) (FE); temperate broadleaf and mixed forest (southern group) (FA); temperate grasslands, savannahs, and shrublands (GN); desert and xeric shrublands (DE); Mediterranean forests, woodlands, and shrub (MS); temperate grasslands, savannahs, and shrublands (GS); tropical and subtropical grasslands, savannahs, and shrublands (GT); temperate coniferous forest (FW). The ecoregions map divides the conterminous United States into eight biomes each representing an assemblage of biophysical, climate, botanical, and animal habitat characteristics defining a distinct geographical area. These biomes were used to stratify sampling of the U.S. cities because climate factors and other biogeographical information are contained in them. Also, it is necessary to understand the dynamic arena within which ecological processes and anthropogenic influences such as urbanization most strongly interact. Totally, 38 of the most populated urban areas in the conterminous United States occurring across six of the largest biomes were analyzed. These urban areas are therefore grouped based on their biomes.

Individual urban settings were identified using the impervious surface area data from the Landsat-derived, continental-scale NLCD to stratify them internally according to ISA density, and estimate their size. They found that a 25% ISA threshold could draw a reasonable boundary between urban and low-intensity residential lands and provide reasonably spatially coherent urban groupings. Therefore, the 25% threshold was used to define polygons in the Landsat-based thematic data. It is worth to mention that the polygons defined here overlap named cities but do not necessarily match incorporated or administrative boundaries. Another important factor is the close match between these urban polygons and the MODIS land cover map classifying urban built-up land. MODIS LST retrieval algorithms estimate surface emissivity based on this land cover map, so a close match here ensures that temperature comparisons within the urban polygon are based on retrievals using the same parameter sets. After urban polygons are defined in individual city place, the landscape within and around them is further stratified into five zones using classes based on ISA and distance. These zones based on classes of percent ISA in concentric rings emanating outward from the highest ISA in a city to the lowest: (1) urban core = pixels having from 75% to 100% ISA; (2) urban in medium density = pixels having ISA between 75% and 50%; (3) urban in low density = pixels having between 50% and 25% ISA—this is the last urban zone and its outer boundary identifies the 25% threshold; (4) suburban = pixels located in a buffer zone 0–5 km adjacent to and outside the 25% ISA contour; and (5) rural = pixels located in a 5 km wide ring located between 45 and 50 km away from the 25% ISA contour and having less than 5% ISA. This rural ring is chosen to be at an optimal distance far enough from the urban core to represent a remote rural area yet not too far to infringe into the 25% contour of an adjacent urban area or another biome. In addition, areas that fall into overlapping biomes, other urban areas, or topographic elevations ±50 m off the mean elevation of the urban core are excluded from the analysis.

The surface temperature and the vegetation within the ISA zones are characterized using MODIS-Aqua eight-day composite (MOD11A2) LST with high quality control and 16-day composite NDVI at 1×1 km^2 resolution. The average summer daytime MODIS LST (1:30 PM local time averaged for June, July, and August) for all cities and zones show that temperatures increase with ISA in all cases except urban areas in desert and xeric shrubland ecoregions, showing an increase outward from the urban core in all biomes. The anomalous NDVI and LST patterns for desert and xeric shrubland cities is likely a result of increased vegetation and latent heat flux in less dense 50%–25% urban and suburban fringe areas due to resource (water) augmentation in those areas. This pattern has been noted previously for U.S. desert cities using AVHRR (Imhoff et al. 2004) and Landsat (Xian and Crane 2006).

The UHI effect is analyzed by calculating the average temperature differences (Urban Core LST–Rural LST) for all the cities in each biome for summer (June/July/August) and winter (December/January/February) daytime (1:30 PM) and nighttime (1:30 AM). The UHI responses for the eight different biomes were all significantly different ($p = 0.01$) and can be used to characterize the effect of ecological context on seasonal and diurnal UHI amplitudes.

The mean intensity of UHI for different biomes suggests that surface temperatures have the greatest differences in daytime. On average, the summer UHI is significantly larger than the winter UHI. The intensity of the summer daytime UHI appears to be related to the standing biomass of the surrounding biome decreasing from forests to grasslands and reversing to a heat sink in desert cities. The largest average summer daytime UHI (7°C–9°C), for example, is exhibited from cities displacing temperate broadleaf, mixed, and temperate coniferous forest biomes (FE, FA, and FW). In high biomass forested biomes, dense and tall vegetation around urban areas intercepts and re-evaporates precipitation and diffuses water from the deep soil to the atmosphere during the process of photosynthesis, thus resulting in the surrounding regions much cooler than the less vegetated urban core. In contrast, urban areas dominated by shorter low biomass vegetation such as grassland, shrublands, and savannah (groups GN, MS, GS, and GT) are shown a less intense UHI with amplitudes ranging from 4°C to 6°C. In these regions, the nonurban core zones are sparsely vegetated and restore an important part of the absorbed solar energy as sensible heating; thus, they reduce the horizontal temperature gradient between the urban core and the rural zone.

However, in urban areas surrounded by deserts and xeric shrublands (DE), this temperature contrast is relatively weak or even is in a reversed pattern. The summer daytime UHI data for DE was actually slightly negative (–1°C) and taken in isolation, it would tend to corroborate the heat sink effect noted for many desert cities. However, the summer nighttime and winter daytime and nighttime UHIs for DE are still positive and large enough to make the cities warmer overall than the surrounding areas.

Furthermore, urban areas in the DE biome are the only ones that consistently exhibit a larger nighttime UHI effect. In their study, average summer daytime LSTs across ISA zones in two metropolitan areas of Baltimore, Maryland, located in the northeastern temperate broadleaf and mixed forest (group FE) and Las Vegas, Nevada, an urban area built within the desert and xeric shrubland biome and arid climate conditions, are also compared to examine the influence of ecological context. The two urban areas have similar average fractional ISA in each of defined urban zone; however, the density of the vegetation in surrounding areas is apparently different. Vegetation density defined by NDVI is large in the urban core of Baltimore and is small in Las Vegas. The spatial variation of NDVI in Las Vegas follows the similar pattern as illustrated by Xian and Crane (2006): increasing away from urban core and dropping again in rural area. Following the ecological contexts, the urban core in Baltimore exhibits a well-defined UHI with an amplitude of 9.3°C, while it shows a possible heat sink (about 0.5°C) in Las Vegas.

To analyze the relationship between the amplitude of UHI and the total size of the urban area, the total area of each urban polygon, which is summed from areas of zones ISA 25% and greater, to the urban core and rural temperature is compared for numbers of cities within the NE biome. Generally, the summer daytime UHI is strongly correlated to the size of the urban area.

The averaged temperatures in urban areas are about 1.5°C warmer than rural areas for urban sizes smaller than 10 km². In the winter daytime, this UHI pattern still exists but with a much weaker UHI effect. The temperature difference could vary from 2°C for urban areas smaller than 10 km² to 3.5°C for urban sizes larger than 1000 km². Their analysis also demonstrated a similar pattern during the nighttime. The difference between the summer and winter UHI amplitudes reveals the vegetation function in these temperate mixed forests. During summer when vegetation is physiologically active, the evaporative cooling is strong during the day and a pronounced UHI between impervious and vegetated zones is produced. During winter, the contrast between the urban and rural zones is subdued when leaves are off and photosynthetic activity is downregulated by cold temperatures. During the nighttime, the UHI pattern remains the same as in daytime but with relatively smaller temperature contrast between urban and rural areas. During summer nighttime, the UHI effects are observed in all urban areas but with smaller magnitudes. The temperature difference between urban and rural areas vary from 1°C for urban areas smaller than 10 km² to 3.5°C for urban areas larger than 1000 km². The UHI effect is less than 1°C for the largest urban area during winter nights. In the Northeastern temperate broadleaf mixed forest biome, they demonstrated a relationship between the UHI magnitude and size of urban area by a log–linear equation:

$$\Delta T_{\text{urban–rural}} = 3.48 \log(\text{area}) + 1.75$$

This formula represents 71% of the variance in UHI with a standard error of ±1.6°C. This equation is also similar to the log–linear relationship between the UHI and the population size developed by Oke (1973, 1976) and Landsberg (1981) except that the equation developed here is directly related to the size of the developed surface (i.e., total area in a city having ISA ≥ 25%). The relationship between temperature difference and impervious surface also suggests that ISA can be used as an objective estimator of the extent and intensity of urban land cover. Also, land surface temperature derived from remote sensing data can be used to produce an accurate estimate of both the magnitude and spatial extent of the UHI effects around the world.

9.3.4 Regional UHI Effects in Europe

A different approach from a general analysis of global UHI effect illustrated in Section 9.2.3 was introduced by focusing on specific UHI effects from a total of 263 European cities (Schwarz et al. 2011). These cities have provided spatial data on their administrative boundaries for 2001 plus basic statistical data, such as population number, and are located on the European continent. City extents are delineated against using administrative boundaries. Furthermore, the MODIS land cover product has been used to quantify indicators such as difference between urban and water. However, the indicator difference between core urban and rural areas was identified using administrative boundaries. The images provided by MODIS on both Terra and Aqua satellites were used. The LST products of MOD11A2 from Terra and of MYD11A2 from Aqua were chosen with a resolution of approximately 1000 m and aggregated for eight consecutive days to calculate the monthly mean LST. LST data for July 2002 and January 2003 were used to represent the summer and winter surface UHI, respectively.

The difference of surface temperature between urban and rural areas is quantified by estimating the indicator difference between core and rural from using the mean LST of administrative city minus the mean LST of a 20 km buffer around the city. The indicator difference of urban and other pixels accounts for temperature differences between urban land cover and other presumably rural land within the administrative area. Furthermore, several additional indicators, such as urban and water and urban and agricultural land, are also included here. The analysis was conducted through several procedures. First, differences in the LST between urbanized land and other land cover types, such as the difference between urban and agriculture or difference between urban and water are included. Second, surface temperatures are interpreted as a Gaussian bell form that can be measured in terms of height when the increasing temperature and ground area are included in the estimation. Third, the total area with temperatures higher than one standard deviation above the mean is used as an indicator for hot island area. Fourth, the magnitude of difference between maximum and mean of the LST in the administrative area is calculated. Finally, the micro-UHI is calculated by estimating percentage of the area with a surface temperature higher than the warmest temperatures associated with tree canopies. Additionally, two simple indicators were added, including the standard deviation of the LSTs in a city and the height of the Gaussian bell that was measured empirically. For most of the indicators, the surface UHI exhibited highest trends in daytime (13:30 h) and lowest at nighttime (01:30 h or 22:30 h). Only the indicator of hot island area showed the opposite tendency of a smaller surface UHI during the day, although only a few points in time were statistically significant. Furthermore, the difference between core and rural showed a smaller strength of surface UHI in Kelvin with a lower diurnal change.

Their analysis also suggested that the surface UHI measured during daytime was not strongly correlated with the surface UHI measured at night for the same indicator for measurements in both summer and winter across European cities. The changes of UHI between July 2002 and 2003 show statistical differences between these two months, suggesting that the surface UHI was significantly larger for July 2002 in daytime, when the surface UHI was largest during the diurnal cycle for most indicators. On the contrary, the surface UHI was larger in July 2003 than in July 2002 at night. The mean surface UHI per indicator and month, however, exhibits that differences for urban-agriculture, urban-other, Gaussian magnitude, magnitude, and Gaussian magnitude empirical are statistically significant between July 2002 and 2003. For all indicators, the monthly surface UHI was larger for July 2002 than for

July 2003. However, a comparison of the mean LST for all cities between July 2002 and July 2003 showed that there were higher temperatures in 2003 than in the previous year ($p < .01$). The warm temperature in 2003 represents a severe heat wave occurred in Europe after a warm and dry spring in parts of the continent (Robine et al. 2008; Dousset et al. 2011).

The study for UHI in European cities suggested that the surface UHI quantified by a single indicator throughout the day and across seasons was not strongly correlated over time. Accordingly, one city that has a high surface UHI during daytime compared to all other cities might have a lower surface UHI at night. A comparison of the surface UHI among a large set of diverse cities might lead to different rankings for different areas in time. Other factors, such as topographical as well as climate characteristics and even local weather conditions, likely influence the temporal dynamics of the surface UHI. Also, the selection of indicators to measure the surface UHI is important. Only a few indicators showed strong correlations for a single location in time, suggesting that specific indicators should be selected for such comparisons in order to compare the surface UHI effects on the same city or on different cities. The recommendation for indicator selection includes the following: (1) only one indicator out of an indicator group with strong relationships is suggested, for example, difference for urban-other (omitting difference for urban-agriculture to also include urban areas with only small agricultural parts); magnitude (omitting standard deviation, Gaussian magnitude, and Gaussian magnitude empirical); and Gaussian area (omitting hot island area); (2) the indicators difference between core and rural and between micro and UHI are more closely reliant to the specific spatial delineation of the urban area; (3) the indicator difference between urban and water is only available for cities with large water bodies, and it shows low stability and should be avoided at least for cities without large water bodies.

9.4 Global Aspect of UHI Effect

MODIS LST data have been used to quantify UHI effects around the world. An early effort for global UHI study focused on using MODIS product of surface temperature, surface emissivity, and albedo over urban areas. The global latitudinal distribution urban nocturnal surface temperature and diurnal range were inspected for cities that are concentrated between 30° and 65°N in the Northern Hemisphere and a few cities in the Tropics and in midlatitudes in the Southern Hemisphere (Jin et al. 2005). Significant longitudinal variations in urban surface temperature are observed associated with the north–south distribution of net radiation received at Earth's surface. The largest range at 30°N of more than 30°C occurs in areas where desert occupies the largest fraction of land. Similarly, the diurnal range over the same latitude shows substantial difference, largely determined by the local or regional sky conditions and surrounding land cover condition. On average, the urban surface temperatures are about 1°C–5°C higher than those over croplands, with larger differences during nighttime in the Southern (winter) Hemisphere. However, the daytime surface temperature differences between the urban areas and adjacent croplands do not have apparent difference but still detectable, especially in the ranges of 10°S–40°N and 30°–45°S. Cities at high latitudes in the Northern Hemisphere (e.g., 55°–60°N) are even cooler than the rural areas during daytime. Large surface temperature differences are also notable at the desert latitudes of 15°–25°N and 10°–20°S because surface temperatures in desert urban areas are close to those in surrounding deserts but higher than those in nearby croplands. A large drop in urban surface temperature takes place near 15°N where few cities exist. The amplitude of diurnal surface temperature range is small (<10°C) in the Tropics and Southern Hemisphere, where oceanic influences are the most pronounced. However, such variation range is large (>30°C) at middle to high latitudes in the Northern Hemisphere. Furthermore, the differences between the urban and forest regions over subtropical and middle latitudes of the Northern Hemisphere are about 4°C with extremes up to 12°C at around 22°N. Urban differs more from forests during the day than it does at night.

The global UHI analysis using MODIS data also found that urban construction reduces both albedo and emissivity (Jin et al. 2005). Magnitudes of urban albedo are about 2%–5% smaller than those in the adjacent croplands. The largest urban albedos are observed over the desert areas around 30°N, which are accompanied by high daytime and nighttime surface temperatures. However, surface emissivity is about 1%–2% lower than that of near croplands.

9.5 Climate Impacts of UHI

The urban built environment is a significant forcing function on the weather-climate system (Seto et al. 2009). Urban *heat islands* are an extreme case of how land use modifies regional climate. UHI could modify the urban climate and such modification is spatially correlated with regional land use and land use change. The reduced vegetation cover, increased in impervious surface area and morphology of buildings in cityscapes, combine to lower evaporative cooling, store heat, and warm the surface air (Foley et al. 2005). The analysis of climate records in the United States suggests that a major portion of the temperature increase during the last several decades resulted from urbanization and other land use changes (Kalnay and Cai 2003). Urbanization and other land use changes accounted for half of the observed reduction in diurnal temperature range and an increase in mean air temperature of 0.27°C in the continental United States during the last century. By comparison, downtown temperatures for the United States have increased by 0.14°C–1.1°C per summertime of cooler regions). The UHI is a local phenomenon with negligible effect on global climate, but its magnitude and effects may represent indications of future climates, as already-observed temperature increases within cities exceed the predicted rise in global temperature for the next several decades (Grimm et al. 2008). Remote sensing-derived UHI information provides both spatial extent and intensity of urban land cover change and its impact on local and regional temperature. Cities offer real-world laboratories to understand these fundamental patterns and processes and to work with city planners, engineers, and architects to implement policies that maximize and sustain livable environments.

The UHI effect may also reduce precipitation by changing surface properties, such as vegetation cover, roughness, albedo, energy flows, and water flows in ways that reduce water supplies to the local atmosphere. Structures associated with urban areas also may change surface hydrology in ways that accelerate runoff via storm water management, which would reduce surface storage and ultimately the water that is available for evaporation. This effect may be exacerbated by a reduction in vegetative cover, which would slow the transfer of water from the soil to the atmosphere via evapotranspiration. This notion is supported by empirical analyses that indicate urbanization reduces the fraction of net radiation that is used for evaporative processes (Carlson and Arthur 2000; Arthur-Hartranft et al. 2003). A statistical analysis of the relationship between climate and urban land use, derived from Landsat data, in concentric buffers around the climate record stations indicates a causal relationship between the temporal and spatial patterns of urbanization and the temporal and spatial patterns of precipitation during the dry season (Kaufmann et al. 2007). Urbanization reduces local precipitation and this reduction may be caused by changes in surface hydrology that extend beyond the UHI effect and energy-related aerosol emissions.

A recent study conducted in Dallas, Texas, using measurements from urban meteorological stations with the rural area stations in the perimeter of the region, as well as by using infrared satellite data, analyzed that UHI and its relation to the 2011 severe drought in the region (Winguth and Kelp 2013). The canopy heat island of a large metropolitan area with a humid, subtropical climate estimated by MODIS data and surface observation shows that the maximum temperature differences between urban and rural stations occurred after sunset, and reached a peak of 3.4°C in July 2011, during a particularly severe heat wave and drought. The drought also produced a *cool* island observed during morning hours after sunrise in response to a slower warming of urban surfaces. On the other hand, high soil moisture and cloud cover have a damping effect on the heat island.

9.6 Summary

This chapter has given an overview of remote sensing techniques and research interests for LST detection and monitoring. UHI and its climatic effects have been analyzed. Major satellite data that have been used to study UHI effects are introduced. The long-term observation and large spatial cover of AVHRR data are still useful for quantifying urban land cover and surface temperature change. However, the spatial resolution of AVHRR is too coarse for most UHI investigations. Landsat imagery has been used for characterizing land cover and surface temperature. It is a very useful data source for local and regional UHI-related studies. The focus of research has been surrounding the UHI effect and the quantification of the relationship between remotely sensed surface temperature and its impact on local and regional climate condition. A significant advantage of remote sensing data and techniques is their global coverage and continuous observation to the entire Earth. However, there are not enough studies focusing on many geographical areas, and a limited number that integrate additional ground data. Remote sensing techniques offer access to data that would otherwise be unobtainable; therefore, the requirement for defensible verification and accuracy measurements is practicable. Together with this, the increasing need for data and intensifying analysis will request using remote sensing data associated with other datasets from numerous sources, resulting in an integral role for remote sensing techniques within a wide range of user communities, including environmental planning, public health, air quality, and urban planning.

References

Chapter 1

Alberti, M., Marzluff, J.M., Shulenberger, E., Bradley, G., Ryan, C., and Zumbrunnen, C. (2003). Integrating humans into ecosystems: Opportunities and challenges for urban ecology. *BioScience*, 53(12), 1169–1179.

Arnfield, A.J. (2003). Two decades of urban climate research: A review of turbulence, exchanges of energy and water, and urban heat island. *International Journal of Climatology*, 23, 1–26.

Carlson, T.N. (2004). Analysis and prediction of surface runoff on an urbanizing watershed using satellite imagery. *Journal of the American Water Resources Association*, 40, 1087–1098.

Gago, E.J., Roldan, J., Pacheco-Torres, R., and Ordóñez, J. (2013). The city and urban heat islands: A review of strategies to mitigate adverse effects. *Renewable and Sustainable Energy Reviews*, 25, 749–758.

Grimmond, S. (2007). Urbanization and global environmental change: Local effects of urban warming. *The Geographical Journal*, 173, 83–88.

Gurjar, B.R., Butler, T.M., Lawrence, M.G., and Lelieveld, J. (2008). Evaluation of emissions and air quality in megacities. *Atmospheric Environment*, 42, 1593–1606.

Hebbert, M., and Jankovic, V. (2013). Cities and climate change: The precedents and why they matter. *Urban Studies*, 50, 1332–1347.

Imhoff, M.L., Bounoua, L., DeFries, R., Lawrence, W.T., Stutzer, D., Tucker, C.J., and Ricketts, T. (2004). The consequences of urban land transformation on net primary productivity in the United States. *Remote Sensing of Environment*, 89, 434–443.

IPCC (2007). *Climate Change 2007: The Physical Science Basis*. Contribution of Working Group I to the Fourth Assessment Report of the Intergovernmental Panel on Climate Change, Cambridge University Press, Cambridge, p. 996.

Jacobson, C.R. (2011). Identification and quantification of the hydrological impacts of imperviousness in urban catchments: A review. *Journal of Environmental Management*, 92, 1438–1448.

Janković, V. (2013). A historical review of urban climatology and the atmospheres of the industrialized world. *Wiley Interdisciplinary Reviews: Climate Change*, 4, 539–553.

Jensen, J.R., and Cowen, D.C. (1999). Remote sensing of urban/suburban infrastructure and socio-economic attributes. *Photogrammetric Engineering and Remote Sensing*, 65, 611–622.

Kanakidou, M. et al. (2011). Megacities as hot spots of air pollution in the East Mediterranean. *Atmospheric Environment*, 45, 1223–1235.

Kukla, G., Gavin, J., and Karl, T.R. (1986). Urban warming. *Journal of Climate and Applied Meteorology*, 25, 1265–1270.

Landsberg, E.H. (1981). *The Urban Climate*. International Geophysics Series, Vol. 28. New York: Academic Press, p. 5.

Lawrence, M.G., Butler, T.M., Steinkamp, J., Gurjar, B.R., and Lelieveld, J. (2007). Regional pollution potentials of megacities and other major population centers. *Atmospheric Chemistry and Physics*, 7, 3969–3987.

Matthews, M.W. (2011). A current review of empirical procedures of remote sensing in inland and near-coastal transitional waters. *International Journal of Remote Sensing*, 32, 6855–6899.

Miller, R.B., and Small, C. (2003). Cities from space: Potential applications of remote sensing in urban environmental research and policy. *Environmental Science & Policy*, 6, 129–137.

Nichol, J., and Wong, M.S. (2005). Modeling urban environmental quality in a tropic city. *Landscape Urban Plan*, 73, 49–58.

Oke, T.R. (1973). City size and the urban heat island. *Atmospheric Environment*, 7, 769–779.

Patino, J.E., and Duque, J.C. (2013). A review of regional science applications of satellite remote sensing in urban settings. *Computers, Environment and Urban Systems*, 37, 1–17.

Quattrochi, D.A., Luvall, J.C., Richman, D.L., Estes, M.G., Laymon, C.A., and Howell, B.F. (2000). A decision support information system for urban landscape management using thermal infrared data. *Photogrammetric Engineering Remote Sensing*, 66, 1195–1207.

Ren, C., Ng, E.Y.-Y., and Katzschner, L. (2011). Urban climate map studies: A review. *International Journal of Climatology*, 31, 2213–2233.

Revl, A., Satterthwaite, D.E., Aragón-Durand, F., Corfee-Morlot, J., Kiunsi, R.B.R., Pelling, M., Roberts, D.C., and Solecki, W. (2014). Urban areas. In: *Climate Change 2014: Impacts, Adaptation, and Vulnerability. Part A–Global and Sectoral Aspects*, Contribution of Working Group II to the Fifth Assessment Report of the Intergovernmental Panel on Climate Change, Field, C.B. et al. (Eds.). Cambridge University Press, Cambridge, pp. 535–612.

Ritchie, J.C., Zimba, P.V., and Everitt, J.H. (2003). Remote sensing techniques to assess water quality. *Photogrammetric Engineering and Remote Sensing*, 69, pp. 695–704.

Schneider, A., Friedl, M.A., and Potere, D. (2009). A new map of global urban extent from MODIS satellite data. *Environmental Research Letters*, 4, 044003, doi:10.1088/1748-9326/4/4/044003.

Seto, K.C., and Shepherd, J.M. (2009). Global urban land-use trends and climate impacts. *Current Opinion in Environmental Sustainability*, 1, 89–95.

Seto, K.C., Güneralp, B., and Hutyra, L. (2012). Global forecasts of urban expansion to 2030 and direct impacts on biodiversity and carbon pools. *Proceedings of the National Academy of Sciences of the United States of America*, 109(40), 16083–16088.

Shepherd, J.M. (2005). A review of current investigations of urban-induced rainfall and recommendations for the future. *Earth Interactions*, 9(1), 1–27.

UNFPA (2007). *The State of World Population 2007: Unleashing the Potential of Urban Growth*. United Nations Population Fund, United Nations Publications, New York, 1 pp.

United Nations (2012). *World Urbanization Prospects. The 2011 Revision*. United Nations, New York, 32 pp.

Voogt, J.A., and Oke, T.R. (2003). Thermal remote sensing of urban climates. *Remote Sensing Environment*, 86, 370–384.

Weng, Q. (2001). A remote-sensing GIS evaluation of urban expansion and its impact on surface temperature in the Zhujiang Delta, China. *International Journal of Remote Sensing*, 22, 1999–2014.

Wulder, M.A., Masek, J.G., Cohen, W.B., Loveland, T.R., and Woodcock, C.E. (2012). Opening the archive: How free data has enabled the science and monitoring promise of Landsat. *Remote Sensing of Environment*, 122, 2–10.

Xian, G. (2007). An analysis of impacts of urban land use and land cover on air quality in the Las Vegas region using remote sensing information and ground observations. *International Journal of Remote Sensing*, 28(24), 5427–5445.

Chapter 2

Agüera, F., and Liu, J.G. (2009). Automatic greenhouse delineation from QuickBird and IKONOS satellite images. *Computers and Electronics in Agriculture*, 66, 191–200.

Aguilar, M.A., Saldaña, M.M., and Aguilar, F.J. (2013). GeoEye-1 and WorldView-2 pan-sharpened imagery for object-based classification in urban environment. *International Journal of Remote Sensing*, 34, 2583–2606.

Blaschke, T., and Strobl, J. (2001). What's wrong with pixels? Some recent developments interfacing remote sensing and GIS. *Journal for Spatial Information and Decision Making*, 14, 12–17.

De Pinho, C.M.D., Fonseca, L.M.G., Korting, T.S., de Almeida, C.M., and Kux, H.J.H. (2012). Land-cover classification of an intra-urban environment using high-resolution images and object-based image analysis. *International Journal of Remote Sensing*, 33(19), 5973–5995.

Drăgut, L., Tiede, D., and Levick, S.R. (2010). ESP: A tool to estimate scale parameter for multiresolution image segmentation of remotely sensed data. *International Journal of Geographical Information Science*, 24, 859–871.

Gamba, P., Dell'Acqua, F., Stasolla, M., Trianni, G., and Lisini, G. (2011). Limits and challenges of optical very-high-spatial-resolution satellite remote sensing for urban applications. In: *Urban Remote Sensing*, Yang, X. (Ed.). Wiley-Blackwell, Hoboken, NJ, 388pp.

Herold, M., Liu, X., and Clarke, K.C. (2003). Spatial metrics and image texture for mapping urban land use. *Photogrammetric Engineering and Remote Sensing*, 69, 991–1001.

Hofmann, P., Strobl, J., Blaschke, T., and Kux, H. (2008). Detecting informal settlements from QuickBird data in Rio de Janeiro using an object-based approach. In: *Object-Based Image Analysis: Spatial Concepts for Knowledge-Driven Remote Sensing Applications*, Blaschke, T., Lang, S., and Hay, G.J. (Eds.). Springer-Verlag, Berlin, Germany, pp. 531–554.

Johansen, K., Arroyo, L.A., and Phinn, S. (2010). Comparison of geo-object based and pixel-based change detection of riparian environments using high spatial resolution multispectral imagery. *Photogrammetric Engineering and Remote Sensing*, 76, 123–136.

Marpu, P.R., Neubert, M., Herold, H., and Niemeyer, I. (2010). Enhanced evaluation of image segmentation results. *Journal of Spatial Science*, 55, 55–68.

Mathieu, R., Aryal, J., and Chong, A.K. (2007). Object-based classification of IKONOS imagery for mapping large-scale vegetation communities in urban areas. *Sensors*, 7, 2860–2880.

Miller, R.B., and Small, C. (2003). Cities from space: Potential applications of remote sensing in urban environmental research and policy. *Environmental Science & Policy*, 6(2), 129–137.

Myint, S.W., Gober, P., Brazel, A., Grossman-Clarke, S., and Weng, Q. (2011). Perpixel vs. object-based classification of urban land cover extraction using high spatial resolution imagery. *Remote Sensing of Environment*, 115, 1145–1161.

Patino, J.E., and Duque, J.C. (2013). A review of regional science applications of satellite remote sensing in urban settings. *Computers, Environment and Urban Systems*, 37, 1–17.

Roelfsema, C., Phinn, S., Jupiter, S., Comley, J., Beger, M., and Paterson, E. (2010). The application of object based analysis of high spatial resolution imagery for mapping large coral reef systems in the West Pacific at geomorphic and benthic community spatial scales, *IEEE International Geoscience and Remote Sensing Society Symposium*, July 25–30, Honolulu, HI. IEEE International Geoscience and Remote Sensing Society, Hawaii, pp. 4346–4349.

Small, C. (2005). A global analysis of urban reflectance. *International Journal of Remote Sensing*, 26(4), 661–681.

Thanapura, P., Helder, D.L., Burckhard, B., Warmath, E., O'Neill, M., and Galster, D. (2007). Mapping urban land cover using quickbird NDVI and GIS spatial modeling for runoff coefficient determination. *Photogrammetric Engineering & Remote Sensing*, 73(1), 57–65.

Thomas, N., Hendrix, C., and Congalton, R.G. (2003). A comparison of urban mapping methods using high-resolution digital imagery. *Photogrammetric Engineering & Remote Sensing*, 69, 963–972.

USGS (2002). Landsat Program Report 2002. U.S. Geological Survey. (https://landsat.usgs.gov/documents/Second_Landsat_Program_Report_FY_2002.pdf).

Weng, Q., and Hu, X. (2008). Medium spatial resolution satellite imagery for estimating and mapping urban impervious surfaces using LSMA and ANN. *IEEE Transactions on Geosciences and Remote Sensing*, 46, 2397–2406.

Xian, G. (2007). Assessing urban growth with subpixel impervious surface coverage. In: *Urban Remote Sensing*, Weng, Q., and Quattrochi, D.A. (Eds.). CRC Press, Boca Raton, FL, pp. 412.

Xian, G. (2008). Mapping impervious surfaces using classification and regression tree algorithm. In: *Remote Sensing of Impervious Surfaces*, Weng, Q. (Eds.). CRC Press, Boca Raton, FL, pp. 441.

Chapter 3

Adams, J.B., Smith, M.O., and Johnson, P.E. (1986). Spectral mixture modeling: A new analysis of road and soil types at the Viking Lander site. *Journal of Geophysical Research*, 91, 8098–8112.

Atkinson, P.M., and Tatnall, A.R.L. (1997). Neural networks in remote sensing. *International Journal of Remote Sensing*, 18, 699–709.

Cadenasso, M.L., Pickett, S.T.A., and Schwarz, K. (2007). Spatial heterogeneity in urban ecosystems: Reconceptualizing land cover and a framework for classification. *Front Ecology Environment*, 5(2), 80–88.

Civco, D.L. (1993). Artificial neural networks for land-cover classification and mapping. *International Journal of Geographical Information Systems*, 7(2), 173–186.

Cristianini, N., and Shawe-Taylor, J. (2000). *An Introduction to Support Vector Machines and Other Kernel-Based Learning Methods*. Cambridge University Press, Cambridge, p. 189.

Fausett, L. (1994). *Fundamentals of Neural Networks: Architectures, Algorithms, and Applications*. Prentice Hall, Englewood Cliffs, NJ, p. 461.

Foody, G.M., McCulloch, M.B., and Yates, W.B. (1995). Classification of remotely sensed data by an artificial neural network: Issues related to training data characteristics. *Photogrammetric Engineering and Remote Sensing*, 61, 391–401.

Green, A.A., Berman, M., Switzer, P., and Craig, M.D. (1988). A transformation for ordering multispectral data in terms of image quality with implications for noise removal. *IEEE Transactions on Geoscience and Remote Sensing*, 26, 65–74.

Homer, C., Huang, C., Yang, L., Wylie, B., and Coan, M. (2004). Development of a 2001 national land cover database for the United States. *Photogrammetric Engineering & Remote Sensing*, 70, 829–840.

Lu, D., and Weng, Q. (2004). Spectral mixture analysis of the urban landscape in Indianapolis city with Landsat ETM+ imagery. *Photogrammetric Engineering & Remote Sensing*, 70, 1053–1062.

Madhavan, B.B., Kubo, S., Kurisaki, N., and Sivakumar, T.V.L.N. (2001). Appraising the anatomy and spatial growth of the Bangkok Metropolitan area using a vegetation impervious-soil model through remote sensing. *International Journal of Remote Sensing*, 22, 789–806.

NLCD (2011). National Land Cover Database. http://www.mrlc.gov/nlcd2011.php (accessed October 1, 2014).

Paola, J.D., and Schowengerdt, R.A. (1995). A review and analysis of back propagation neural networks for classification of remotely sensed multispectral imagery. *International Journal of Remote Sensing*, 16, 3033–3058.

Pickett, S.T.A., Burch, W.R., and Dalton, S. (1997). Integrated urban ecosystem research. *Urban Ecosystems*, 1, 183–184.

Powell, R., Roberts, D.A., Hess, L., and Dennison, P. (2007). Sub-pixel mapping of urban land cover using multiple endmember spectral mixture analysis: Manaus, Brazil. *Remote Sensing of Environment*, 106(2), 253–267.

Quinlan, J.R. (1993). *C4.5: Programs for Machine Learning*. Morgan Kaufmann, San Francisco, CA.

Rashed, T., Weeks, J.R., Roberts, D., Rogan, J., and Powell, R. (2003). Measuring the physical composition of urban morphology using multiple endmember spectral mixture models. *Photogrammetric Engineering & Remote Sensing*, 69, 1011–1020.

Ridd, M.K. (1995). Exploring a V-I-S (vegetation-impervious surface-soil) model for urban ecosystem analysis through remote sensing: Comparative anatomy for cities. *International Journal of Remote Sensing*, 16, 2165–2185.

Roberts, D.A., Gardner, M., Church, R., Ustin, S., Scheer, G., and Green, R.O. (1998). Mapping chaparral in the Santa Monica mountains using multiple endmember spectral mixture models. *Remote Sensing of Environment*, 65, 267–279.

Roberts, D.A., Smith, M.O., and Adams, J.B. (1993). Green vegetation, nonphotosynthetic vegetation, and soil in AVIRIS data. *Remote Sensing of Environment*, 44, 255–269.

Sabol, D.E., Gillespie, A.R., Adams, J.B., Smith, M.O., and Tucker, C.J. (2002). Structural stage in pacific northwest forests estimated using simple mixing models of multispectral images. *Remote Sensing of Environment*, 80, 1–16.

Small, C. (2001). Estimation of urban vegetation abundance by spectral mixture analysis. *International Remote Sensing*, 22, 1305–1334.

Small, C. (2003). High spatial resolution spectral mixture analysis of urban reflectance. *Remote Sensing of Environment*, 88, 170–186.

Small, C. (2005). A global analysis of urban reflectance. *International Journal of Remote Sensing*, 26, 661–681.

Smith, M.O., Johnson, P.E., and Adams, J.B. (1985). Quantitative determination of mineral types and abundances from reflectance spectra using principal component analysis. *Journal of Geophysical Research*, 90, 792–804.

Smith, M.O., Ustin, S.L., Adams, J.B., and Gillespie, A.R. (1990). Vegetation in deserts: I. A regional measure of abundance from multispectral images. *Remote Sensing of Environment*, 81, 427–442.

Song, C. (2005). Spectral mixture analysis for subpixel vegetation fractions in the urban environment: How to incorporate endmember variability? *Remote Sensing of Environment*, 95, 248–263.

Strahler, A.H., Woodcock, C.E., and Smith, J.A. (1986). On the nature of models in remote sensing. *Remote Sensing of Environment*, 70, 121–139.

Sun, Z., Guo, H., Li, X., Lu, L., and Dua, X. (2011). Estimating urban impervious surfaces from Landsat-5 TM imagery using multilayer perceptron neural network and support vector machine. *Journal of Applied Remote Sensing*, 5, 53501–53517.

Vapnik, V.N. (2000). *The Nature of Statistical Learning Theory*, 2nd edition. Springer-Verlag, New York, p. 314.

Walton, J.T. (2008). Subpixel urban land cover estimation: Comparing cubist, random forests, and support vector regression. *Photogrammetric Engineering & Remote Sensing*, 74, 1213–1222.

Ward, D., Phinn, S.R., and Murray, A.T. (2000). Monitoring growth in rapidly urbanizing areas using remotely sensed data. *The Professional Geographer*, 53, 371–386.

Weng, Q. (2012). Remote sensing of impervious surfaces in the urban areas: Requirements, methods, and trends. *Remote Sensing of Environment*, 117, 34–49.

Weng, Q., and Hu, X. (2008). Medium spatial resolution satellite imagery for estimating and mapping urban impervious surfaces using LSMA and ANN. *IEEE Transactions on Geoscience and Remote Sensing*, 46, 2397–2406.

Weng, Q., and Lu, D. (2007). Subpixel analysis of urban landscapes. In: *Urban Remote Sensing*, Weng, Q., and Quattrochi, D.A. (Eds.). CRC Press, Boca Raton, FL, p. 405.

Weng, Q., and Lu, D. (2009). Landscape as a continuum: An examination of the urban landscape structures and dynamics of Indianapolis City, 1991–2000, by using satellite image. *International Journal of Remote Sensing*, 30, 2547–2577.

Wu, C., and Murray, A.T. (2003). Estimating impervious surface distribution by spectral mixture analysis. *Remote Sensing of Environment*, 84(4), 493–505.

Xian, G. (2007). Assessing urban growth with sub-pixel impervious surface coverage. In: *Urban Remote Sensing*, Weng, Q., and Quattrochi, D.A. (Eds.). Taylor & Francis Group, Boca Raton, FL, pp. 179–199.

Xian, G. (2008). Mapping impervious surfaces using classification and regression tree algorithm. In: *Remote Sensing of Impervious Surfaces*, Weng, Q. (Ed.). Taylor & Francis Group, Boca Raton, FL, pp. 39–58.

Xian, G., and Crane, M. (2005). Assessments of urban growth in the Tampa Bay watershed using remote sensing data, *Remote Sensing of Environment*, 97, 203–205.

Xian, G., and Homer, C. (2010). Updating the 2001 national land cover database impervious surface products to 2006 using Landsat imagery change detection methods. *Remote Sensing of Environment*, 114, 1676–1686.

Xian, G., Homer, C., Dewitz, J., Fry, J., Hossain, N., and Wickham, J. (2011). Change of impervious surface area between 2001 and 2006 in the conterminous United States. *Photogrammetric Engineering & Remote Sensing*, 77(8), 758–762.

Xian, G., Yang, L., Klaver, J.M., and Hossain, N. (2006). Measuring urban sprawl and extent through multi-temporal imperviousness mapping. *Rates, Trends, Causes, and Consequences of Urban Land-Use Change in the United States*, USGS Professional Paper 1762, 59–64.

Yang, L., Xian, G., Klaver, J.M., and Deal, B. (2003). Urban land-cover change detection through sub-pixel imperviousness mapping using remotely sensed data, *Photogrammetric Engineering and Remote Sensing*, 69, 1003–1010.

Yang, X., and Zhou, L. (2011). Parameterizing neural network models to improve land classification performance. In: *Urban Remote Sensing: Monitoring, Synthesis and Modeling in the Urban Environment*, Yang, X. (Ed.), Wiley-Blackwell, Hoboken, NJ, p. 383.

Chapter 4

Bartholome, E., and Belward, A.S. (2005). GLC2000: A new approach to global land cover mapping from Earth observation data. *International Journal of Remote Sensing*, 26, 1959–1977.

Belward, A.S., and Loveland, T. (1997). The IGBP-DIS global 1 km land cover data set, DISCover: First results. *International Journal of Remote Sensing*, 18, 3291–3295.

Cakir, H.I., Khorram, S., and Nelson, S.A.C. (2006). Correspondence analysis for detecting land cover change. *Remote Sensing of Environment*, 102, 306–317.

CIESIN (Center for International Earth Science Information Network) (2004). Global Rural–Urban Mapping Project (GRUMP), Alpha Version: Urban Extents. http:// sedac.ciesin.columbia.edu/gpw (accessed September 1, 2013).

Croft, T. (1978). Nighttime images of the earth from space. *Scientific American*, 239, 86–98.

Danko, D.M. (1992). The digital chart of the world project. *Photogrammetric Engineering and Remote Sensing*, 58, 1125–1128.

DeFries, R.S., Hansen, M.C., Townshend, J.R.G., Janetos, A.C., and Loveland, T.R. (2000). A new global 1-km dataset of percentage tree cover derived from remote sensing. *Global Change Biology*, 6, 247–254.

Elvidge, C., Imhoff, M.L., Baugh, K.E., Hobson, V.R.., Nelson, I., Safran, J., Dietz, J.B., and Tuttle, B.T. (2001). Nighttime lights of the world: 1994–95. *ISPRS Journal of Photogrammetry and Remote Sensing*, 56, 81–99.

Elvidge, C., Tuttle, B., Sutton, P., Baugh, K., Howard, A., Milesi, C., Bhaduri, B., and Nemani, R. (2007a). Global distribution and density of constructed impervious surfaces. *Sensors*, 7, 1962–1979.

Elvidge, C.D. et al. (2007b). The nightsat mission concept. *International Journal of Remote Sensing*, 28, 2645–2670.

Esch, T., Himmler, V., Schorcht, G., Thiel, M., Wehrmann, T., Bachofer, F., Conrad, C., Schmidt, M., and Dech, S. (2009). Large-area assessment of impervious surface based on integrated analysis of single-date Landsat-7 images and geospatial vector data. *Remote Sensing of Environment*, 113, 1678–1690.

Esch, T., Thiel, M., Schenk, A., Roth, A., Mehl, H., and Dech, S. (2010). Delineation of urban footprints from TerraSAR-X data by analyzing speckle characteristics and intensity information. *IEEE Transactions on Geoscience and Remote Sensing*, 48(2), 905–916.

Friedl, M., and Brodley, C. (1997). Decision tree classification of land cover from remotely sensed data. *Remote Sensing of Environment*, 61, 399–409.

Friedl, M.A. et al. (2002). Global land cover mapping from MODIS: Algorithms and early results. *Remote Sensing of Environment*, 83, 287–302.

Friedl, M.A., Sulla-Menashe, D., Tan, B., Schneider, A., Ramankutty, N., Sibley, A., and Huang, X. (2009). MODIS Collection 5 Global Land Cover: Algorithm refinements and characterization of new datasets. *Remote Sensing of Environment*, 114, 168–182.

Grimm, N.B., Grove, J.M., Pickett, S.T.A., and Redman, C.L. (2000). Integrated approaches to long-term studies of urban ecological systems. *BioScience*, 50, 571–584.

Goldewijk, K. (2001). Estimating global land use change over the past 300 years: The HYDE database. *Global Biogeochemical Cycles*, 15, 417–434.

Goldewijk, K. (2005). Three centuries of global population growth: A spatially referenced population density database for 1700–2000. *Population and Environment*, 26, 343–367.

Hansen, M., Dubayah, R., and DeFries, R. (1996). Classification trees: An alternative to traditional land cover classifiers. *International Journal of Remote Sensing*, 17, 1075–1081.

Hansen, M.C., Defries, R.S., Townsend, R.G., and Sohlberg, R. (2000). Global land cover classification at 1 km spatial resolution using a classification tree approach. *International Journal of Remote Sensing*, 21, 1331–1364.

Homer, C., Dewitz, J., Fry, J., Coan, M., Hossain, N., Larson, C., Herold, N., McKerrow, A., VanDriel, J.N., and Wickham, J. (2007). Completion of the 2001 National Land Cover Database for the conterminous United States. *Photogrammetric Engineering & Remote Sensing*, 73, 337–341.

Homer, C., Huang, C., Yang, L., Wylie, B., and Coan, M. (2004). Development of a 2001 national land cover database for the United States. *Photogrammetric Engineering & Remote Sensing*, 70, 829–840.

Huber, M., Wessel, B., and Roth, A. (2006). The TerraSAR-X orthorectification service and its benefit for land use applications. *IEEE International Geoscience and Remote Sensing Symposium*, Denver, CO.

Huete, A., Didan, K., Miura, T., Rodriguez, E.P., Gao, X., and Ferreira, L.G. (2002). Overview of the radiometric and biophysical performance of the MODIS vegetation indices. *Remote Sensing of Environment*, 83, 195–213.

Imhoff, M.L., Lawrence, W.T., Elvidge, C.D., Paul, T., Levine, E., Privalsky, M.V., and Brown, V. (1997). Using nighttime DMSP/OLS images of city lights to estimate the impact of urban land use on soil resources in the United States. *Remote Sensing of Environment*, 59, 105–117.

Loveland, T.R., Reed, B.C., Brown, J.F., Ohlen, D.O., Zhu, J., Yang, L., and Merchant, J.W. (2000). Development of a global land cover characteristics database and

IGBP DISCover from 1-km AVHRR data. *International Journal of Remote Sensing*, 21, 1303–1330.

Loveland, T.R., Sohl, T.L., Stehman, S.L.V., Gallant, A.L., Sayler, K.L., and Napton, D.E. (2002). A strategy for estimating the rates of recent United States land-cover changes. *Photogrammetric Engineering & Remote Sensing*, 68, 1091–1099.

McCallum, I., Obersteiner, M., Nilsson, S., and Shvidenko, A. (2006). A spatial comparison of four satellite derived 1 km global land cover datasets. *International Journal of Applied Earth Observation and Geoinformation*, 8, 246–255.

Pal, M., and Mather, P.M. (2003). An assessment of the effectiveness of decision tree methods for land cover classification. *Remote Sensing of Environment*, 86, 554–565.

Pickett, S.T.A. et al. (2008). Beyond urban legends: An emerging framework of urban ecology, as illustrated by the Baltimore Ecosystem Study. *BioScience*, 58, 139–150.

Pickett, S.T.A., Cadenasso, M.L., Grove, J.M., Nilon, C.H., Pouyat, R.V., Zipperer, W.C., and Costanza, R. (2001). Urban ecological systems: Linking terrestrial ecological, physical, and socioeconomic components of metropolitan areas. *Annual Review of Ecology and Systematics*, 32, 127–157.

Potere, D., and Schneider, A. (2007). A critical look at representations of urban areas in global maps. *GeoJournal*, 69, 55–80.

Potere, D., Schneider, A., Schlomo, A., and Civco, D.A. (2009). Mapping urban areas on a global scale: Which of the eight maps now available is more accurate? *International Journal of Remote Sensing*, 30, 6531–6558.

Quinlan, J.R. (1996). Bagging, boosting and C4.5 AAAI-96. *Proceedings of the 13th National Conference on Artificial Intelligence*, Portland, OR, August. AAAI Press, Menlo Park, CA, pp. 725–730.

Robert, C.P. (1997). *The Bayesian Choice: A Decision-Theoretic Motivation*. Springer-Verlag, New York, 436 pp.

Schaaf, C.B. et al. (2002). First operational BRDF, albedo nadir reflectance products from MODIS. *Remote Sensing of the Environment*, 83, 135–148.

Schneider, A., Friedl, M.A., Mciver, D.K., and Woodcock, C.E. (2003). Mapping urban areas by fusing multiple sources of coarse resolution remotely sensed data. *Photogrammetric Engineering & Remote Sensing*, 69, 1377–1386.

Schneider, A., Friedl, M.A., and Potere, D. (2009). A new map of global urban extent from MODIS satellite data. *Environmental Research Letters*, 4, 044003. doi:10.1088/1748-9326/4/4/044003.

Schneider, A., Friedl, M.A., and Potere, D. (2010). Mapping global urban areas using MODIS 500-m data: New methods and datasets based on "urban ecoregions." *Remote Sensing of Environment*, 114, 1733–1746.

Schneider, A., and Woodcock, C.E. (2008). Compact, dispersed, fragmented, extensive? A comparison of urban expansion in twenty-five global cities using remotely sensed data, pattern metrics and census information. *Urban Studies*, 45, 659–692.

Seto, K.C., and Fragkias, M. (2005). Quantifying spatiotemporal patterns of urban landuse change in four cities of China with a time series of landscape metrics. *Landscape Ecology*, 20, 871–888.

Small, C. (2003). High spatial resolution spectral mixture analysis of urban reflectance. *Remote Sensing of Environment*, 88, 170–186.

Small, C. (2005). A global analysis of urban reflectance. *International Journal of Remote Sensing*, 26, 661–681.

Taubenböck, H. (2008). Vulnerabilitätsabschätzung der Megacity Istanbul mit Methoden der Fernerkundung. PhD thesis. University of Würzburg, Würzburg, Germany, p. 178.

Taubenböck, H., Esch, T., Felbier, A., Wisener, M., Roth, A., and Dech, S. (2012). Monitoring urbanization in mega cities from space. *Remote Sensing of Environment*, 117, 162–176.

Vogelmann, J.E., Howard, S.M., Yang, L., Larson, C.R., Wylie, B.K., and Van Driel, N. (2001). Completion of the 1990's National Land Cover Data Set for the conterminous United States from Landsat Thematic Mapper Data and ancillary data sources. *Photogrammetric Engineering & Remote Sensing*, 67, 650–662.

Weng, Q. (2012). Remote sensing of impervious surfaces in the urban areas: Requirements, methods, and trends. *Remote Sensing of Environment*, 117, 34–49.

Xian, G., and Homer, C. (2010). Updating the 2001 National Land Cover Database impervious surface products to 2006 using Landsat imagery change detection methods. *Remote Sensing of Environment*, 114(2), 1676–1686.

Xian, G., Homer, C., Bunde, B., Danielson, P., Dewitz, J., Fry, J., and Pu, R. (2012). Quantifying urban land cover change between 2001 and 2006 in the Gulf of Mexico region. *Geocarto International*, 27(6), 479–497.

Xian, G., Homer, C., Dewitz, J., Fry, J., Hossain, N., and Wickham, J. (2011). Change of impervious surface area between 2001 and 2006 in the conterminous United States. *Photogrammetric Engineering & Remote Sensing*, 77(8), 758–762.

Xian, G., Homer, C., and Fry, J. (2009). Updating the 2001 National Land Cover Database land cover classification to 2006 by using Landsat imagery and change detection methods. *Remote Sensing of Environment*, 113, 1133–1147.

Chapter 5

Arnold Jr., C.A., and Gibbons, C.J. (1996). Impervious surface coverage: The emergence of a key urban environmental indicator. *Journal of the American Planning Association*, 62(2), 243–258.

Babin, M., and Stramski, D. (2002). Light absorption by aquatic particles in the near-infrared spectral region. *Limnology and Oceanography*, 47, 911–915.

Babin, M., Stramski, D., Ferrari, G.M., Claustre, H., Bricaud, A., Obolensky, G., and Hoepffner, N. (2003). Variations in the light absorption coefficients of phytoplankton, nonalgal particles, and dissolved organic matter in coastal waters around Europe. *Journal of Geophysical Research*, 108, 3211–3230.

Brivio, P.A., Giardino, C., and Zilioli, E. (2001). Determination of chlorophyll concentration changes in Lake Garda using an image-based radiative transfer code for Landsat TM images. *International Journal of Remote Sensing*, 22, 487–502.

Caraco, N.F. (1995). Influence of human populations on P transfers to aquatic systems: A regional scale study using large rivers. In: *Phosphorus in the Global Environment. SCOPE 54*, Tiessen, H. (Ed.). Wiley, New York, pp. 235–247.

Cavalli, R.M., Laneve, G., Fusilli, L., Pignatti, S., and Santini, F. (2009). Remote sensing water observation for supporting Lake Victoria weed management. *Journal of Environment Management*, 90, 2199–2211.

Dekker, A. (1993). Detection of optical water quality parameters for eutrophic waters by high resolution remote sensing. PhD thesis, Free University, Amsterdam, the Netherlands.

Dekker, A.G., Vos, R.J., and Peters, S.W.M. (2001). Comparison of remote sensing data, model results and in situ data for the Southern Frisian Lakes. *Science of the Total Environment*, 268, 197–214.

Dekker, A.G., Vos, R.J., and Peters, S.W.M. (2002). Analytical algorithms for lake water TSM estimation for retrospective analyses of TM and SPOT sensor data. *International Journal of Remote Sensing*, 23, 15–35.

Doxaran, D., Froidefond, J.M., Castaing, P., and Barin, M. (2009). Dynamics of the turbidity maximum zone in a macrotidal estuary (the Gironde, France): Observations from field and MODIS satellite data. *Estuarine, Coastal and Shelf Science*, 81, 321–332.

Giardino, C., Candiani, G., and Zilioli, E. (2005). Detecting chlorophyll-a in Lake Garda using TOA MERIS radiances. *Photogrammetric Engineering and Remote Sensing*, 71, 1045–1051.

Gibbons, D.E., G.E. Wukelic, Leighton, J.P., and Doyle, M.J. (1989). Application of Landsat Thematic Mapper data for coastal thermal plume analysis at Diablo Canyon. *Photogrammetric Engineering & Remote Sensing*, 55, 903–909.

Gitelson, A. (1992). The peak near 700 nm on radiance spectra of algae and water: Relationships of its magnitude and position with chlorophyll concentration. *International Journal of Remote Sensing*, 13, 3367–3373.

Gitelson, A., Garbuzov, G., Szilagyi, F., Mittenzwey, K., Karnieli, A., and Kaiser, A. (1993). Quantitative remote sensing methods for real-time monitoring of inland waters quality. *International Journal of Remote Sensing*, 14, 1269–1295.

Gitelson, A.A., Gritz, Y., and Merzlyak, M.N. (2003). Relationships between leaf chlorophyll content and spectral reflectance and algorithms for non-destructive chlorophyll assessment in higher plant leaves. *Journal of Plant Physiology*, 160, 271–282.

Gitelson, A.A., Gurlin, D., Moses, W.J., and Barrow, T. (2009). A bio-optical algorithm for the remote estimation of the chlorophyll-a concentration in case 2 waters. *Environmental Research Letters*, 4, 45003.

Gove, N.E., Edwards, R.T., and Conquest, L.L. (2001). Effects of scale on land use and water quality relationships: A longitudinal basin-wide perspective. *Journal of the American Water Resource Association*, 37, 1721–1734.

Gower, J.F.R., Doerffer, R., and Borstad, G.A. (1999). Interpretation of the 685 nm peak in water-leaving radiance spectra in terms of fluorescence, absorption and scattering, and its observation by MERIS. *International Journal of Remote Sensing*, 20, 1771–1786.

Harding, L.W., Itsweire, E.C., and Esaias, W.E. (1995). Algorithm development for recovering chlorophyll concentrations in the Chesapeake Bay using aircraft remote sensing, 1989–1991. *Photogrammetric Engineering & Remote Sensing*, 61, 177–185.

Hillsborough County (1999). Watershed Management Plan. http://www.hillsborough.wateratlas.usf.edu (accessed on August 2, 2005).

Hu, C. (2009). A novel ocean color index to detect floating algae in the global oceans. *Remote Sensing of Environment*, 113, 2118–2129.

Hu, C., Lee, Z., Ma, R., Yu, K., Li, D., and Shang, S. (2010). Moderate resolution imaging spectroradiometer (MODIS) observations of cyanobacteria blooms in Taihu Lake, China. *Journal of Geophysical Research*, 115, 1–2.

Kutser, T., Arst, H., Mäekivi, S., and Kallaste, K. (1998). Estimation of the water quality of the Baltic Sea and lakes in Estonia and Finland by passive optical remote sensing measurements on board vessel. Lakes and Reservoirs. *Research and Management*, 3, 53–66.

Kutser, T., Pierson, D.C., Kallio, K.Y., Reinart, A., and Sobek, S. (2005). Mapping lake CDOM by satellite remote sensing. *Remote Sensing of Environment*, 94, 535–540.

Matthews, M.W. (2011). A current review of empirical procedures of remote sensing in inland and near-coastal transitional waters. *International Journal of Remote Sensing*, 32, 6855–6899.

Matthews, M.W., Bernard, S., and Robertson, L. (2012). An algorithm for detecting trophic status (chlorophyll-a), cyanobacterial-dominance, surface scums and floating vegetation in inland and coastal waters. *Remote Sensing of Environment*, 124, 637–652.

Matthews, M.W., Bernard, S., and Winter, K. (2010). Remote sensing of cyanobacteria-dominant algal blooms and water quality parameters in Zeekoevlei, a small hypertrophic lake, using MERIS. *Remote Sensing of Environment*, 114, 2070–2208.

Moses, W.J., Gitelson, A.A., Berdnikov, S., and Povazhnyy, V. (2009). Satellite estimation of chlorophyll-a concentration using the red and NIR bands of MERIS—The Azov Sea case study. *IEEE Geoscience and Remote Sensing Letters*, 6, 845–849.

Oki, K., and Yasuoka, Y. (2008). Mapping the potential annual total nitrogen load in the river basins of Japan with remote-sensing imagery. *Remote Sensing of Environment*, 112, 3091–3098.

Olmanson, L.G., Bauer, M.E., and Brezonik P.L. (2008). A 20-year Landsat water clarity census of Minnesota's 10,000 lakes. *Remote Sensing of Environment*, 112, 4086–4097.

Olmanson, L.G., Brezonik, P.L., and Bauer, M.E. (2011). Evaluation of medium to low resolution satellite imagery for regional lake water quality assessments. *Water Resources Research*, 47, W09515. doi:10.1029/2011WR011005.

Onderka, M., and Pekarova, P. (2008). Retrieval of suspended particulate matter concentrations in the Danube River from Landsat ETM data. *Science of the Total Environment*, 397, 238–243.

O'Reilly, J.E. et al. (1998). Ocean color chlorophyll algorithms for SeaWiFS. *Journal of Geophysical Research*, 103, 24937–24953.

Oyama, Y., Matsushita, B., Fulkushima, K., Matsushige, T., and Imai, A. (2009). Application of spectral decomposition algorithm for mapping water quality in a turbid lake (Lake Kasumigaura, Japan) from Landsat TM data. *ISPRS Journal of Photogrammetric Engineering & Remote Sensing*, 64, 73–85.

Peng, G. (2012). Remote sensing of environmental change over China: A review. *Chinese Science*, 57, 2793–2801.

Ritchie, J.C., Schiebe, F.R., and McHenry, J.R. (1976). Remote sensing of suspended sediment in surface water. *Photogrammetric Engineering & Remote Sensing*, 42, 1539–1545.

Ritchie, J.C., Zimba, P.V., and Everitt, J.H. (2003). Remote sensing techniques to assess water quality. *Photogrammetric Engineering & Remote Sensing*, 69, 695–704.

Ruiz-Verdú, A., Simis, S.G.H., DE Hoyos, C., Gons, H.J., and Peña-Martinea, R. (2008). An evaluation of algorithms for the remote sensing of cyanobacterial biomass. *Remote Sensing of Environment*, 112, 3996–4008.

Schalles, J.F., Gitelson, A.A., Yacobi, Y.Z., and Kroenke, A.E. (1997). Estimation of chlorophyll a from time series measurements of high spectral resolution reflectance in an eutrophic lake. *Journal of Phycology*, 34, 383–390.

Schiebe, F.R., Harrington, Jr., J.A., and Ritchie, J.C. (1992). Remote sensing of suspended sediments: The Lake Chicot, Arkansas project. *International Journal of Remote Sensing*, 13, 1487–1509.

Schubel, J.P., and Pritchard, D.W. (1986). Responses of upper Chesapeake Bay to variations in discharge of the Susquehanna River. *Estuaries*, 9, 236–249.

Schueler, T.R. (1994). The importance of imperviousness. *Watershed Protection Techniques*, 1(3), 100–111.

Slonecker, E.T., Jennings, D.B., and Garofalo, D. (2001). Remote sensing of impervious surface: A review. *Remote Sensing Review*, 20, 227–235.

Song, S.H., and Wang, M. (2012). Water properties in Chesapeake Bay from MODIS-Aqua measurements. *Remote Sensing of Environment*, 123, 163–174.

USEPA (2001). *Our Built and Natural Environments: A Technical Review of the Interactions between Land Use, Transportation, and Environmental Quality*. US Environmental Protection Agency: Development, Community, and Environment, Washington, DC.

Volpe, V., Silvestri, S., and Marani, M. (2011). Remote-sensing retrieval of suspended sediment concentration in shallow waters. *Remote Sensing of Environment*, 115, 44–54.

Wang, M., Son, S., and Harding, L.W. (2009). Retrieval of diffuse attenuation coefficient in the Chesapeake Bay and turbid ocean regions for satellite ocean color applications. *Journal of Geophysical Research*, 114, C10011. doi:10.1029/2009JC005286.

Wang, M., and Shi, W. (2008). Satellite observed algae blooms in China's Lake Taihu. *EOS Transactions of the American Geophysical Union*, 89, 201–202. doi:10.1029/2008EO220001

Wang, M., Shi, W., and Tang, J. (2011). Water property monitoring and assessment for China's inland Lake Taihu from MODIS-Aqua measurements. *Remote Sensing of Environment*, 115, 841–854.

World Resources Institute (2003). *A Guide to the Global Environment—Environmental Change and Human Health*. The World Resources Institute, The United Nations Environmental Program, The United Nations Development Program, and The World Bank, New York, pp. 1–369.

Xian, G., and Crane, M. (2005). Assessments of urban growth in the Tampa Bay watershed using remote sensing data. *Remote Sensing of Environment*, 97(2), 203–215.

Xian, G., Crane, M., and Su, J. (2007). An analysis of urban development and its environmental impact on the Tampa Bay watershed. *Journal of Environmental Management*, 85, 965–976.

Yacobi, Y.Z., Gitelson, A., and Mayo, M. (1995). Remote sensing of chlorophyll in Lake Kinneret using high spectral-resolution radiometer and Landsat TM: Spectral features of reflectance and algorithm development. *Journal of Plankton Research*, 17, 2155–2217.

Chapter 6

Al-Khudhairy, D.H.A., Caravaggi, I., and Giada, S. (2005). Structural damage assessments from Ikonos data using change detection, object-oriented segmentation, and classification techniques. *Photogrammetric Engineering & Remote Sensing*, 71, 825–837.

Booth, D.B. (1987). Timing and processes of deglaciation along the southern margin of Gordilleran ice sheet. In: *North America and Adjacent Oceans during the Last Deglaciation: The Geology of North America*, Ruddiman W.F., and Wright, H.J. (Eds.). Geological Society of America, Boulder, CO, pp. 71–90.

Church, P.E. (1974). Some precipitation characteristics of Seattle. *Weatherwise*, December, pp. 244–251.

Coe, J.A., Michael, J.A., Crovelli, R.A., Savage, W.Z., Laprade, W.T., and Nashem, W.D. (2004). Probabilistic assessment of precipitation-triggered landslides using historical records of landslide occurrence, Seattle, WA. *Environmental & Engineering*, X(2), 103–122.

Crowell, M., Coulton, K., Johnson, C., Westcott, J., Bellomo, D., Edelman, S., and Hirsch, E. (2010). An estimate of the U.S. population living in 100-year coastal flood hazard areas. *Journal of Coastal Research*, 26, 201–211.

Dai, F., Lee, C., and Ngai, Y. (2002). Landslide risk assessment and management: An overview. *Engineering Geology*, 64, 65–87.

Ehrlich, D., and Tenerelli, P. (2013). Optical satellite imagery for quantifying spatio-temporal dimension of physical exposure in disaster risk assessments. *Natural Hazards*, 68, 1271–1289.

Giustarini, L., Hostache, R., Matgen, P., Schumann, G.J.-P., Bates, P.D., and Mason, D.C. (2013). A change detection approach to flood mapping in urban areas using TerraSAR-X. *IEEE Transactions on Geoscience and Remote Sensing*, 51, 2417–2439.

Godt, J.W., Baum, R.L., and Chlebord, A.F. (2006). Rainfall characteristics for shallow landsliding in Seattle, Washington, USA. *Earth Surface Processes and Landforms*, 31, 97–110.

Grestel, W.J., Brunengo, M.J., Lingley, W.S., Jr., Logan, R.L., Shipman, H., and Walsh, T.J. (1997). Puget Sound Bluffs: The where, why, and when of landslides following the holiday 1996/97 storms. *Washington Geology*, 25(1), 17–31.

Guzzetti, F., Stark, C., and Salvati, P. (2005). Evaluation of flood and landslide risk to the population of Italy. *Environmental Management*, 36(1), 15–36.

Islam, M.M., and Sado, K. (2002). Development priority map for flood countermeasures by remote sensing data with geographic information system. *Journal of Hydrologic Engineering*, 7, 346–355.

Iwan, W.D., Cluff, L.S., Kimpel, J.F., Kunreuther, H., Masakischatz, S.H., Nigg, J.M., Roth, R.S., Sr., Stanley, E., and Thomas, F.H. (1999). Mitigation emerges as major strategy for reducing losses caused by natural disasters. *Science*, 284, 1943–1947.

Joyce, K.E., Bellis, S.E., Samsonov, S.V., McNeill, S.J., and Glassey, P.J. (2009). A review of the status of satellite remote sensing and image processing techniques for mapping natural hazards and disasters. *Progress in Physical Geography*, 33, 183–207.

KCPSB (King County Performance, Strategy, and Budget) (2010). *The Annual Growth Report 2000*. King County, WA. http://www.kingcounty.gov/exec/PSB/Demographics/DataReports.aspx (accessed on August 10, 2014).

Laprade, W.T., Kirkland, T.E., Nashem, W.D., and Robertson, C.A. (2000). *Seattle Landslide Study. Internal Report W-7992-01*. Shannon and Wilson, Seattle, WA, p. 164.

Lorente, A., Garcia-Ruiz, J., Begueria, S., and Arnáez, J. (2002). Factors explaining the spatial distribution of hillslope debris flows. *Mountain Research and Development*, 22, 32–39.

Mantovani, F., Soeters, R., and van Westen, C. (1996). Remote sensing techniques for landslide studies and hazard zonation in Europe. *Geomorphology*, 15, 213–225.

Mason, D.C., Speck, R., Devereux, B., Schumann, G., Neal, J., and Bates, P.D. (2010). Flood detection in urban areas using TerraSAR-X. *IEEE Transactions on Geoscience and Remote Sensing*, 48, 882–894.

McBean, G., and Ajibade, I. (2009). Climate change, related hazards and human settlements. *Current Option in Environmental Sustainability*, 1, 179–186.

McGuirk, J.P. (1982). A century of precipitation variability along the Pacific coast of North America and its impact. *Climatic Change*, 4, 41–56.

Metternicht, G., Hurni, L., and Gogu, R. (2005). Remote sensing of landslides: An analysis of the potential contribution to geo-spatial systems for hazard assessment in mountainous environments. *Remote Sensing of Environment*, 98, 284–303.

Nichol, J., and Wong, M.S. (2005a). Detection and interpretation of landslides using satellite images. *Land Degradation and Development*, 16, 243–255.

Nichol, J., and Wong, M.S. (2005b). Satellite remote sensing for detailed landslide inventories using change detection and image fusion. *International Journal of Remote Sensing*, 26, 1913–26.

Paegeler, M. (1998). *Landslide Policies for Seattle: A Report to the Seattle City Council from the Landslide Policy Group*. Seattle, WA, p. 35.

Robinson, L., Newell, J.P., and Marzluff, J.M. (2005). Twenty-five year of sprawl in the Seattle region: Growth management responses and implication for conservation. *Landscape and Urban Planning*, 71, 51–72.

Schott, J.R. (1997). *Remote Sensing the Image Chain Approach*. Oxford University Press, New York, p. 394.

Schueler, T.R. (1994). The importance of imperviousness. *Watershed Protection Techniques*, 1(3), 100–111.

Seto, K.C., and Shepherd, J.M. (2009). Global urban land-use trends and climate impacts. *Current Opinion in Environmental Sustainability*, 1, 89–95.

Shannon and Wilson (2000). *Geotechnical Report, Feasibility Study: Evaluation of Works Progress Administration Subsurface Drainage Systems, Seattle, Washington*. Internal Report, 21-1-08831-001, Shannon & Wilson, Seattle, WA, pp. 1–41.

Sidle, R.C., and Ochiai, H. (2006). *Landslides, Processes, Prediction, and Land Use*. American Geophysical Union, Washington, DC, p. 312.

Taubenböck, H., Wurm, M., Netzband, M., Zwenzer, H., Roth, A., Rahman, A., and Dech, S. (2011). Flood risks in urbanized areas—multi-sensoral approaches using remotely sensed data for risk assessment. *Natural Hazards and Earth System Science*, 11, 431–444.

Xian, G., and Homer, C. (2013). Assessment of urbanization patterns and trends in the Gulf of Mexico region of the southeast United States with Landsat and night-time lights imagery. In: *Advances in Mapping from Remote Sensor Imagery*, Yang, X., and Li, J. (Eds.). CRC Press, Boca Raton, FL, pp. 185–202.

Xian, G., Homer, C., Bunde, B., Danielson, P., Dewitz, J., Fry, J., and Pu, R. (2012). Quantifying urban land cover change between 2001 and 2006 in the Gulf of Mexico region. *Geocarto International*, 27(6), 479–497.

Chapter 7

Aumann, H.H. et al. (2003). AIRS/AMSU/HSB on the aqua mission: Design, science objectives, data products, and processing systems. *IEEE Transactions on Geoscience and Remote Sensing*, 41, 253–264.

Barnes, W.L., Pagano, T.S., and Salomonson, V.V. (1998). Prelaunch characteristics of the moderate resolution imaging spectroradiometer (MODIS) on EOS-AM1. *IEEE Transactions on Geoscience and Remote Sensing*, 36, 1088–1100.

Bechle, M.J., Millet, D.B., and Mashall, J.D. (2013). Remote sensing of exposure to NO_2: Satellite versus ground-based measurement in a large urban area. *Atmospheric Environment*, 69, 345-353.

Beer, R. (2006). TES on the Aura mission: Scientific objectives, measurements, and analysis overview. *IEEE Transactions on Geoscience and Remote Sensing*, 44, 1102–1105.

Beer, R., Glavich, T.A., and Rider, D.M. (2001). Tropospheric emission spectrometer for the earth observing system's aura satellite. *Applied Optics*, 40, 2356–2367.

Bey, I.D., Jacob, J., Yantosca, R.M., Logan, J.A., Field, B., Fiore, A.M., Li, Q., Liu, H., Mickley, L.J., and Schultz, M. (2001). Global modeling of tropospheric chemistry with assimilated meteorology: Model description and evaluation. *Journal of Geophysical Research*, 106(D19), 23,073–23,096.

Boubel, R., Fox, D.L., Turner, D.B., and Stern, A.C. (1994). *Fundamentals of Air Pollution*, 3rd edition. Academic Press, San Diego, CA.

Bovensmann, H., Burrows, J.P., Buchwitz, M., Frerick, J., Noël, S., Rozanov, V.V., Chance, K.V., and Goede, A.P.H. (1999). SCIAMACHY: Mission objectives and measurement modes. *Journal of the Atmospheric Sciences*, 56, 127–150.

Bucsela, E.J., Celarier, E.A., Wenig, M.O., Gleason, J.F., Veefkind, J.P., Boersma, K.F., and Brinksma, E.J. (2006). Algorithm for NO_2 vertical column retrieval from the ozone monitoring instrument. *IEEE Transactions on Geoscience and Remote Sensing*, 44, 1245–1258.

Burrows, J.P. et al. (1999). The global ozone monitoring experiment (GOME): Mission concept and first scientific results. *Journal of the Atmospheric Sciences*, 56, 151–175.

Callies, J., Corpaccioli, E., Eisinger, M., Hahne, A., and Lefebvre, A. (2000). GOME-2 Metop's second-generation sensor for operational ozone monitoring. *European Space Agency Bulletin*, 102, 28–36.

Cartalis, C., and Varotsos, C. (1994). Surface ozone in Athens, Greece, at beginning and at the end of the twentieth century. *Atmospheric Environment*, 28, 3–8.

Chance, K. (2006). Spectroscopic measurements of tropospheric composition from satellite measurements in the ultraviolet and visible: Steps toward continuous pollution monitoring from space. In: Perrin, A., Ben Sari-Zizi, N., Demaison, J. (Eds.), Remote Sensing of the Atmosphere for Environmental Security. Springer, ISBN 1-4020-5089-5, pp.1-25.

Clerbaux, C. et al. (2009). Monitoring of atmospheric composition using the thermal infrared IASI/MetOp sounder. *Atmospheric Chemistry and Physics*, 9, 6041–6054.

Coutant, B.W., Engel-Cox, J., and Swinton, K.E. (2003). *Compilation of Existing Studies on Source Appointment for $PM_{2.5}$*. Technical Report Draft 1–05. US EPA, Triangle Park, NC.

Deeter, M.N. et al. (2003). Operational carbon monoxide retrieval algorithm and selected results for the MOPITT instrument. *Journal of Geophysical Research*, 108, 4399. doi:10.1029/2002JD003186.

Diner, D.J. et al. (1998). Multi-angle Imaging SpectroRadiometer (MISR) instrument description and experiment overview. *IEEE Transactions on Geoscience and Remote Sensing*, 36, 1072–1087.

Donkelaar, A.V., Martin, R.V., and Park, R.J. (2006). Estimating ground-level PM2.5 using aerosol optical depth determined from satellite remote sensing. *Journal of Geophysical Research*, 111, D21201. doi:10.1029/2005JD006996.

Drummond, J.R. (1992). Measurements of pollution in the troposphere (MOPITT). In: *The Use of EOS for Studies of Atmospheric Physics*, Gille, J.C., and Visconti, G. (Eds.). North-Holland, New York, p. 77e101.

Fishman, J., Vukovich, F.M., Cahoon, D., and Shipman, M.C. (1987). The characterization of an air pollution episode using satellite total ozone measurements. *Journal of Applied Meteorology*, 26, 1638–1654.

Fraser, R.S., Kaufman, Y.J., and Mahoney, R.L. (1984). Satellite measurements of aerosol mass and transport. *Atmospheric Environment*, 18(12), 2577–2584.

Herman, J.R., and Celarier, E.A. (1997). Earth surface reflectivity climatology at 340–380 nm from TOMS data. *Journal of Geophysical Research*, 102, 28003–28011.

Kar, J., Fishman, J., Creilson, J.K., Richter, A., Ziemke, J., and Chandra, S. (2010). Are there urban signatures in the tropospheric ozone column products derived from satellite measurements? *Atmospheric Chemistry and Physics*, 10, 5213–5222.

Kaufman, Y., Fraser, R.S., and Ferrare, R. (1990). Satellite measurements of large-scale air pollution: Methods. *Journal of Geophysical Research*, 95(d7), 9895–9909.

Kaufman, Y.J., Tanré, D., and Boucher, O. (2002). A satellite view of aerosols in the climate system. *Nature*, 419, 215–223.

Kaufman, Y.J., Tanré, D., and Remer, L.A. (1997). Operational remote sensing of tropospheric aerosol over land from EOS moderate resolution imaging spectroradiometer. *Journal of Geophysics Research*, 102, 17051–17067.

Keating, T., and Zhuber, A. (2007). Hemispheric transport of air pollution. In: *Air Pollution Studies* No. 16, United Nations, New York, p. 165.

Klenk, K.F., Bhartia, P.K., Fleig, A.J., Kaveeshwar, V.G., McPeters, R.D., and Smith, P.M. (1982). Total ozone determination from the backscattered ultraviolet (BUV) experiment. *Journal of Applied Meteorology*, 21, 1672–1684.

Kuze, A., Suto, H., Nakajima, M., and Hamazaki, T. (2009). Thermal and near infrared sensor for carbon observation Fourier-transform spectrometer on the Greenhouse Gases Observing Satellite for greenhouse gases monitoring. *Applied Optics*, 48, 6716–6733.

Levelt, P.F., van den Oord, G.H.J., Dobber, M.R., Mälkki, A., Visser, H., de Vries, J., Stammes, P., Lundell, J.O.V., and Saari, H. (2006). The ozone monitoring instrument. *IEEE Transactions on Geoscience and Remote Sensing*, 44, 1093–1101.

Liu, X., Bhartia, P.K., Chance, K., Spurr, R.J.D., and Kurosu, T.P. (2010). Ozone profile retrievals from the ozone monitoring instrument. *Atmospheric Chemistry and Physics*, 10, 2521–2537

Liu, Y., Park, R.J., Jacob, D.J., Li, Q., Kilaru, V., and Sarnat, J.A. (2004a). Mapping annual mean ground-level PM2.5 concentrations using multiangle imaging spectroradiometer aerosol optical thickness over the contiguous United States. *Journal of Geophysical Research*, 109, D22206. doi:10.1029/2004JD005025.

Liu, Y., Sarnat, J.A., Coull, B.A., Koutrakis, P., and Jacob, D.J. (2004b). Validation of multiangle imaging spectroradiometer (MISR) aerosol optical thickness measurements using Aerosol Robotic Network (AERONET) observations over the contiguous United States. *Journal of Geophysical Research*, 109, D06205. doi:10.1029/2003JD003981.

Liu, Y., Sarnat, J.A., Kilaru, V., Jacob, D.J., and Koutrakis, P. (2005). Estimating ground-level $PM_{2.5}$ in the eastern United States using satellite remote sensing. *Environmental Science & Technology*, 39, 3269–3278.

Martin, R.V. (2008). Satellite remote sensing of surface air quality. *Atmospheric Environment*, 42, 7823–7843.

Martin, R.V., Jacob, D.J., Yantosca, R.M., Chin, M., and Ginoux, P. (2003). Global and regional decreases in tropospheric oxidants from photochemical effects of aerosols. *Journal of Geophysical Research*, 108(D3), 4097. doi:10.1029/2002JD002622.

Martin, R.V. et al. (2004). Evaluation of GOME satellite measurements of tropospheric NO_2 and HCHO using regional data from aircraft campaigns in the southeastern United States. *Journal of Geophysical Research*, 109, D24307. doi:10.1029/2004JD004869.

Pan, L., Edwards, D.P., Gille, J.C., Smith, M.W., and Drummond, J.R. (1995). Satellite remote sensing of tropospheric CO and CH_4: Forward model studies of the MOPITT instrument. *Applied Optics*, 34, 6976–6988.

Park, R.J., Jacob, D.J., Chin, M., and Martin, R.V. (2003). Sources of carbonaceous aerosols over the United States and implications for natural visibility. *Journal of Geophysical Research*, 108(D12), 4355. doi:10.1029/2002JD003190.

Pope, C.A., III, Bates, D.V., and Raizenne, M.E. (1995). Health effects of particulate air pollution: Time for reassessment? *Environmental Health Perspectives*, 103, 472–480.

Puett, R.C., Hart, J.E., Yanosky, J.D., Paciorek, C., Schwartz, J., Suh, H., Speizer, F.E., and Laden, F. (2009). Chronic fine and coarse particulate esposure, mortality, and coronary heart disease in the nurses' health study. *Environmental Health Perspectives*, 117, 1697–1701.

Russell, A.R., Valin, L.C., Bucsela, E.J., Wenig, M.O., and Cohen, R.C. (2010). Space-based constraints on spatial and temporal patterns of NO_x emissions in California, 2005–2008. *Environmental Science & Technology*, 44, 3608–3615.

Schoeberl, M.R. et al. (2007). A trajectory-based estimate of the tropospheric ozone column using the residual method. *Journal of Geophysical Research*, 112, D24S49. doi:10.1029/2007JD008773.

Sifakis, N., and Deschamps, P.Y. (1992). Mapping of air pollution using SPOT satellite data. *Photogrammetric Engineering and Remote Sensing*, 58, 1433–1437.

Streets, D.G. et al. (2013). Emissions estimation from satellite retrievals: A review of current capability. *Atmospheric Environment*, 77, 1011–1042.

Todd, W.J., George, A.J., and Bryant, N.A. (1979). Satellite-aided evaluation of population exposure to air pollution. *Environmental Science & Technology* 13, 970–974.

Wang, J., and Christopher, S.A. (2003). Intercomparison between satellite-derived aerosol optical thickness and PM2.5 mass: Implications for air quality studies. *Geophysical Research Letters*, 30(21), 2095. doi:10.1029/2003GL018174.

Wang, Z., Chen, L., Tao, J., Zhang, Y., and Su, L. (2010). Satellite-based estimation of regional particulate matter (PM) in Beijing using vertical-and-RH correcting method. *Remote Sensing of Environment*, 114, 50–63.

World Health Organization (2000). *Air Quality Guidelines for Europe*, 2nd edition. WHO Regional Publications, Copenhagen, Denmark. European series, Chapter 7, vol. 91.

Xian, G. (2007). Analysis of impacts of urban land use and land cover on air quality in the Las Vegas region using remote sensing information and ground observation. *International Journal of Remote Sensing*, 28, 5427–5445.

Zhang, L. et al. (2008). Transpacific ozone pollution mechanisms and effect of recent Asian emission increases on air quality in North America. *Atmospheric Chemistry and Physics*, 8, 8143–8191.

Zhang, L., Jacob, D.J., Liu, X., Logan, J.A., Chance, K., Eldering, A., and Bojkov, B.R. (2010). Intercomparison methods for satellite measurements of atmospheric composition: Application to tropospheric ozone from TES and OMI. *Atmospheric Chemistry and Physics*, 10, 4725–4739.

Zhang, R., Lei, W., Tie, X., and Hess, P. (2004). Industrial emissions cause extreme urban ozone diurnal variability. *Proceedings of the National Academy of Sciences of the United States of America*, 101, 6346–6350.

Chapter 8

Carlson, T.N., and Wendling, P. (1977). Reflected radiance measured by NOAA 3 VHRR as a function of optical depth for Saharan dust. *Journal of Applied Meteorology*, 16, 1368–1371.

Duce, R.A., Unni, C.K., Ray, B.J., Prospero, J.M., and Merrill, J.T. (1980). Long-range atmospheric transport of soil dust from Asia to the tropical North Pacific: Temporal variability. *Science*, 209, 1522–1524.

Fraser, R.S. (1976). Satellite measurements of mass of Sahara dust in the atmosphere. *Applied Optics*, 15, 2471–2479.

Fraser, R.S., Kaufman, Y.J., and Mahoney, R.L. (1984). Satellite measurements of aerosol mass and transport. *Atmospheric Environment*, 18, 2577–2584.

Grimm, N.B., Faeth, S.H., Golubiewski, N.E., Redman, C.L., Wu, J., Bai, X., and Briggs, J.M. (2008). Global change and ecology of cities. *Science*, 319, 756–760.

Gupta, P., Christopher, S.A., Wang, J., Gehrig, R., Lee, Y., and Kumar, N. (2006). Satellite remote sensing of particulate matter and air quality assessment over global cities. *Atmospheric Environment*, 40, 5880–5890.

Holben, B.N. et al. (1998). AERONET—A federated instrument network and data archive for aerosol characterization. *Remote Sensing of Environment*, 66, 1–16.

Kaufman, Y.J., Koren, I., Remer, L.A., Rosenfeld, D., and Rudich, Y. (2005a). The effect of smoke, dust and pollution aerosol on shallow cloud development over the Atlantic Ocean. *Proceedings of the National Academy of Sciences*, 102, 11207–11212.

Kaufman, Y.J., Koren, I., Remer, L.A., Tanré, D., Ginoux, P., and Fan, S. (2005b). Dust transport and deposition observed from the Terra-Moderate Resolution Imaging Spectroradiometer (MODIS) spacecraft over the Atlantic Ocean. *Journal of Geophysical Research*, 110, D10S12. http://dx.doi.org/10.1029/2003JD004436.

Kaufman, Y.J., Tanré, D., . León, J.-F., and Pelon (2003). Retrievals of profiles of fine and coarse aerosols using lidar and radiometric space measurements. *IEEE Trans. Geosci. Remote Sens.*, 41(8), 1743–1754, doi:10.1109/TGRS.2003.814138.

Lee, Y.C., and Hills, P.R. (2003). Cool season pollution episodes in Hong Kong, 1996–2002. *Atmospheric Environment*, 37, 2927–2939.

Martin, R.V. (2008). Satellite remote sensing of surface air quality. *Atmospheric Environment*, 42, 7823–7843.

Mekler, Y., Quenzel, H., Ohring, G., and Marcus, I. (1977). Relative atmospheric aerosol content from ERTS observations. *Journal of Geophysical Research*, 82, 967–970.

Remer, L.A. et al. (2005). The MODIS aerosol algorithm, products and validation. *Journal of the Atmospheric Sciences*, 62, 947–973.

Streets, D.G. et al. (2013). Emissions estimation from satellite retrievals: A review of current capability. *Atmospheric Environment*, 77, 1011–1042.

United States Environmental Protection Agency (2003). *Air Quality Index: A Guide to Air Quality and Your Health*. EPA-454/K-03–002, United States Environmental Protection Agency. http://www.airnow.gov.

Uno, I., Eguchi, K., Yumimoto, K., Takemura, T., Shimizu, A., Uematsu, M., Liu, Z., Wang, Z., Hara, Y., and Sugimoto, N. (2009). Asian dust transported one full circuit around the globe. *Nature Geoscience*, 2, 557–560. http://dx.doi.org/10.1038/ngeo583.

Yu, H. et al. (2013b). An HTAP multi-model assessment of the influence of regional anthropogenic emission reductions on aerosol direct radiative forcing and the role of intercontinental transport. *Journal of Geophysical Research: Atmospheres*, 118. doi:10.1029/2012JD018148.

Yu, H., Chin, M., Winker, D.M., Omar, A., Liu, Z., Kittaka, C., and Diehl, T. (2010). Global view of aerosol vertical distributions from CALIPSO lidar measurements and GOCART simulations: Regional and seasonal variations. *Journal of Geophysical Research*, 115, D00H30. doi:10.1029/2009JD013364.

Yu, H., Quinn, P.K., Feingold, G., Remer, L.A., Kahn, R.A., Chin, M., and Schwartz, S.E. (2009). Remote sensing and in situ measurements of aerosol properties, burdens, and radiative forcing, in atmospheric aerosol properties and climate impacts. In: *A Report by the U.S. Climate Change Science Program and the Subcommittee on Global Change Research*, Chin, M., Kahn, R.A., and Schwartz, S.E. (Eds.). National Aeronautics and Space Administration, Washington, DC.

Yu, H., Remer, L.A., Chin, M., Bian, H., Kleidman, R., and Diehl, T. (2008). A satellite based assessment of trans-Pacific transport of pollution aerosol. *Journal of Geophysical Research*, 113, D14S12. doi:10.1029/2007JD009349.

Yu, H., Remer, L.A., Kahn, R.A., Chin, M., and Zhang, Y. (2013a). Satellite perspective of aerosol intercontinental transport: From qualitative tracking to quantitative characterization. *Atmospheric Research*, 124, 73–100.

Chapter 9

Arnfield, A.J. (2003). Two decades of urban climate research: A review of turbulence, exchanges of energy and water, and the urban heat island. *International Journal of Climatology*, 23, 1–26.

Arthur-Hartranft, T., Carlson, N., and Clarke, K.C. (2003). Satellite and ground-based microclimate and hydrological analyses coupled with a regional urban growth model. *Remote Sensing of Environment*, 86, 385–400.

Carlson, T.N., and Arthur, S.T. (2000). The impact of land use/land cover changes due to urbanization on surface microclimate and hydrology: A satellite perspective. *Global Planet Change*, 25, 49–65.

Carlson, T.N., Dodd, J.K., Benjamin, S.G., and Cooper, J.N. (1981). Remote estimation of surface energy balance, moisture availability and thermal inertia. *Journal of Applied Metalworking*, 20, 67–87.

Coll, C., Caselles, V., Galve, J., Valor, E., Niclos, R., Sanchez, J., and Rivas, R. (2005). Ground measurements for the validation of land surface temperatures derived from AATSR and MODIS data. *Remote Sensing of Environment*, 97, 288–300.

Cotton, W.R., and Pielke, R.A. (1995). *Human Impacts on Weather and Climate*. Cambridge University Press, Cambridge, p. 288.

Dousset, B. et al. (2011). Satellite monitoring of summer heat waves in the Paris metropolitan area. *International Journal of Climatology*, 31, 313–323.

Foley, J.A. et al. (2005). Global consequences of land use. *Science*, 309, 570–574.

Gallo, K., and Xian, G. (2014). Application of spatially gridded temperature and land cover data sets for urban heat island analysis. *Urban Climate*, 8, 1–10.

Grimm, N.B., Faeth, S.H., Golubiewski, N.E, Redman, C.L., Wu, J., Bai, X., and Briggs, J.M. (2008). Global change and ecology of cities. *Science*, 319, 756–760.

Imhoff, M.L., Zhang, P., Wolfe, R.E., and Bounoua, L. (2010). Remote Sensing of the urban heat island effect across biomes in the continental USA. *Remote Sensing of Environment*, 114, 504–513.

Jin, M., Dickinson, R.E., and Zhang, D.-L. (2005). The footprint of urban areas on global climate as characterized by MODIS. *Journal of Climate*, 18, 1551–1565.

Kalnay, E., and Cai, M. (2003). Impact of urbanization and land-use change on climate. *Nature*, 423, 528–531.

Kaufmann, R.K., Seto, K.C., Schneider, A., Liu, Z., Zhou, L., and Wang, W. (2007). Climate response to rapid urban growth: Evidence of a human-induced precipitation deficit. *Journal of Climate*, 20, 2299–2306.

Landsberg, E.H. (1981). *The Urban Climate*. International Geophysics Series, Vol. 28. Academic Press, New York, p. 275.

Menne, M.J., Durre, I., Vose, R.S., Gleason, B.E., and Houston, T.G. (2012). An overview of the global historical climatology network-daily database. *Journal of Atmospheric and Oceanic Technology*, 29, 897–910.

Oke, T.R. (1973). City size and the urban heat island. *Atmospheric Environment*, 7, 769–779.

Oke, T.R. (1976). The distinction between canopy and boundary layer urban heat islands. *Atmosphere*, 14, 268–277.

Olson, D.M. et al. (2001). Terrestrial ecoregions of the world: A new map of life on earth. *BioScience*, 51(11), 933–938.

Price, J.C. (1984). Land surface temperature measurements from the split window channels of the NOAA 7 advanced very high resolution radiometer. *Journal of Geophysical Research*, 89, 7231–7237.

Robine, J.M. et al. (2008). Death toll exceeded 70,000 in Europe during the summer of 2003. *Comptes Rendus Biologies*, 331, 171–178.

Roth, M., Oke, T.R., and Emery, W.J. (1989). Satellite-derived urban heat island for three coastal cities and the utilization of such data in urban climatology. *International Journal of Remote Sensing*, 10, 1699–1720.

Schwarz, N., Lautenback, S., and Seppelt, R. (2011). Exploring indicators for quantifying surface urban heat islands of European cities with MODIS land surface temperatures. *Remote Sensing of Environment*, 115, 3175–3186.

Seto, K.C., and Shepherd, J.M. (2009). Global urban land-use trends and climate impacts. *Current Opinion of Environmental Sustainability*, 1, 89–95.

Streutker, D.R. (2002). A remote sensing study of the urban heat island of Houston, Texas. *International Journal of Remote Sensing*, 23, 2595–2608.

Streutker, D.R. (2003). Satellite-measured growth of the urban heat island of Houston, Texas. *Remote Sensing of Environment*, 85, 282–289.

Taha, H., Douglas, S., and Haney, J. (1997). Mesoscale meteorological and air quality impacts of increased urban albedo and vegetation. *Energy and Buildings*, 25, 169–177.

Thornton, P.E., Running, S.W., and White, M.A. (1997). Generating surfaces of daily meteorology variables over large regions of complex terrain. *Journal of Hydrology*, 190, 214–251.

Thornton, P.E., Thornton, M.M., Mayer, B.W., Wilhelmi, N., Wei, Y., and Cook, R.B. (2012). Daymet: Daily surface weather on a 1 km grid for North America, 1980–2008. Oak Ridge National Laboratory Distributed Active Archive Center, Oak Ridge, Tennessee. doi:10.3334/ORNLDAAC/Daymet_V2. http://daymet.ornl.gov/ (accessed on May 7, 2013).

Tomlinson, C., Chapman, L., Thornes, J.E., and Baker, C. (2011). Remote sensing land surface temperature for meteorology and climatology: A review. *Meteorological Applications*, 18, 296–306.

Voogt, J.A., and Oke, T.R. (2003). Thermal remote sensing of urban climates. *Remote Sensing of Environment*, 86, 370–384.

Wan, Z. (2002). Validation of the land-surface temperature products retrieved from terra moderate resolution imaging Spectroradiometer data. *Remote Sensing of Environment*, 83, 163–180.

Wan, Z. (2008). New refinements and validation of the MODIS land-surface temperature/emissivity products. *Remote Sensing of Environment*, 112, 59–74.

Wan, Z., and Dozier, J. (1996). A generalized split-window algorithm for retrieving land-surface temperature from space. *IEEE Transactions on Geoscience and Remote Sensing*, 34, 892–905.

Wan, Z., Zhang, Y., Zhang, Q., and Li, Z. (2004). Quality assessment and validation of the MODIS global land surface temperature. *International Journal of Remote Sensing*, 25, 261–274.

Weng, Q., Lu, D., and Schubring, J. (2004). Estimation of land surface temperature-vegetation abundance relationship for urban heat island studies. *Remote Sensing of Environment*, 89, 467–483.

Winguth, A.M.E., and Kelp, B. (2013). The urban heat island of the north-central Texas region and its relation to the 2011 server Texas drought. *Journal of Applied Meteorology and Climatology*, 52, 2418–2433.

Xian, G. (2008). Satellite remotely-sensed land surface parameters and their climatic impacts in multi metropolitan regions. *Advances in Space Research*, 41, 1861–1869.

Xian, G., and Crane, M. (2006). An analysis of urban thermal characteristics and associated land cover in Tampa Bay and Las Vegas using Landsat data. *Remote Sensing of Environment*, 104, 147–156.

Index

Note: Locators followed by "*f*" and "*t*" denote figures and tables in the text

A

Absorption by CDOM (aCDOM), 88–89
Advanced very high resolution
　　radiometer (AVHRR), 61,
　　111, 154
　data, 133
　instrument, 171–173
Aerosol optical depth (AOD), 131, 140, 156
　global distributions of, 165*f*
　global maps of, 155*f*
　measure by MODIS and MISR
　　instruments, 142
　measurements of, 159
　relationship between $PM_{2.5}$ and,
　　141–142
　satellite measurements of, 163
　spatial distribution of, 160
Aerosol plume, identification of, 157
Aerosol remote sensing, 140–141
Aerosol Robotic Network
　　(AERONET), 156
Aerosol size
　global distributions of, 166*f*
　spatial variations of, 155*f*
Aerosol type information, 156
Air mass factor (AMF) calculation, 139
Air pollution
　type of, 129
　urban, 129
Air quality, 161
　different cities, 162
　Hong Kong, 162
　Sydney, 162
　urban areas, 141–147, 153–168
　urban land cover, 147–150
Aqua satellite, 136
　near-polar orbits, 174
Areas of interest (AOI), 45
Artificial neural network (ANN), 46
ASTER, satellite images acquired
　　from, 25*f*

Atmospheric aerosols, 129
　formation of, 130
　remote sensing of, 131
Atmospheric infrared sounder
　　(AIRS), 136
Aura satellite, 136–137

B

Back-propagation (BP), 46
Backscattering characteristics of
　　water, 79
Bayes' rule, 69
Beer's law, 138
Best-fit model, 80
Biological oxygen demand (BOD), 91
　annual, 92*f*
Biomes, 182
Brightness temperature, 173

C

California, locations of metropolitan
　　regions in, 145*f*
Cases file, 41
Chesapeake Bay, 86
　land cover, 99*f*
　water properties in, 96–100
Chlorophyll-a (Chl-a), 81
　concentrations, 82*f*, 85–88
Chromophoric DOM, 82
City of Seattle, 118
　landslides, 118, 123*f*
　shaded-relief map of, 123*f*
Classification and regression trees
　　(CART), 54
Climate records in the United States, 188
Climate system components, 169
Cloudaerosol lidar and infrared
　　pathfinder satellite
　　observations (CALIOP), 154

Coastal urban areas, landslide
    assessment in, 114–126
Colored DOM (CDOM), 79,
    88–89
    retrieval of absorption by, 88
    yellow substances, 88
CONUS, 55, 57
Cubist, 36
    output files from, 41

D

Dai Lake, water conditions in, 78*f*
Dallas metropolitan area, 179*f*
Data file, 38, 41
Daymet grid cell temperatures, 181*f*
Defense Meteorological Satellite
    Program (DMSP), 55
Deserts and xeric shrublands
    (DE), 184
Digital elevation model (DEM), 112
Digital number (DN) value, 18, 30
Digital orthophoto quarter
    quadrangles (DOQQs), 14, 41,
    54, 120
    in Tampa Bay, 15*f*
Dissolved organic matter (DOM), 79
DMSP-OLS data, 64
Donut patterns, 58–59
Dust storms, 155, 156
    during summer, 167

E

Earth observation satellite, 8
Earthquake-prone, 107
Endmembers, 30
    to select, 31
Enhanced Thematic Mapper Plus
    (ETM+), 106
Environmental changes, 2
Envisat satellite, 136
ERS-2 satellite, 133
Expert system rule, 18–19

F

First-generation sensors, 8
Flood-affected frequency, 113–114

Flood risks
    hazard rank, 113
    in urban environments, 111–114
Fossil fuel combustion, 129

G

Gaussian bell, 186
Geostationary Operational
    Environmental Satellite
    (GOES), 131, 154
GIS (geographic information
    system), planning data
    in, 116
GlobalCover v2 (GLOBC), 62*t*, 63
Global distributions of
    aerosol optical depth, 165*f*
    aerosol size, 166*f*
Global Historical Climatology Network-
    Daily Database (GHCN), 179*f*,
    180, 182
Global Impervious Surface Area
    (IMPSA), 62*t*, 63
Global Land Cover Characteristics
    database, 61
Global land cover map, 61
Global Land Cover 2000 v1.1 (GLC00),
    61–63, 62*t*
Global ozone monitoring experiment
    (GOME), 132
Global PM$_{2.5}$ distribution,
    characterization of, 160–163
Global rural-urban mapping project
    (GRUMP), 62*t*, 63
Global-scale product, accuracy
    assessment, 70–71
Global urban land cover, classification,
    68–69
Global urban mapping effort
    image classification, 72–75
    multitemporal change
        detection, 75
    remote sensing data, 72
    urban footprint and
        multitemporal change
        product, 75
Global urban maps, 62*t*
Goddard Earth Observing System
    (GEOS-3), 142

Greenhouse gases, 153
Greenhouse gases Observing SATellite
        (GOSAT), 137
Grid cells, 180
Gulf of Mexico region, 108*f*
    land cover transition, 106–111, 110*f*
    urban areas in, 109*f*
    urban development, 106–111

H

Hazard rank (HR), 113
High-density built-up land,
        124–125
High-resolution satellite imagery,
        application of, 9–12
Hillsborough County, 91
History Database of the Global
        Environment v3 (HYDE3),
        62*t*, 63
Hong Kong, 115–116
    air quality in, 162
    landslides in, 116
    location of study area, 116*f*
    Natural Terrain Landslide Inventory
        data, 117
    polluted region, 161
Hurricane-prone, 107

I

IKONOS (satellite), 8, 106
    spatial resolution of, 9
Impervious surface, 11, 13
Impervious surface area (ISA), 53, 55, 57,
        95, 107, 126*f*, 148
    estimation, 41
    group, 175
    Landsat images, 121*f*
    magnitudes of, 125
    spatial distributions of, 121*f*
Industrial nonpoint pollutants, 91
International Geosphere–Biosphere
        Programme (IGBP), 67–68

J

Japan Aerospace Exploration Agency
        (JAXA), 137

L

Lake Balaton, spectral reflectance
        measure, 82*f*
Lake Chicot, suspended sediments in,
        95–96
Lake Taihu
    in China, 100
    land cover map of, 101*f*
    water quality assessment for, 100–102
Land-clearing fires, 156
Land cover
    data, 55
    types, 13
Land cover and land use (LCLU), 52, 55
Land remote sensing, 8
Landsat, 8
    satellite images acquired from, 25*f*
Landsat data, 72–73
Landsat imagery, 173–174
    thematic mapper, 173
Landsat images, 54, 56, 72, 78*f*
Landsat multispectral scanner (MSS)
        sensor, 25
Landsat systems, 26*t*, 27*t*
Landsat Thematic Mapper (TM), 8,
        53, 106
    data, 25, 89
Landscape, 2
Landslides
    City of Seattle, 118, 123*f*
    evolution of, 122–126
    historical record of, 122
    Hong Kong, 115–116, 116*f*
    records between 1900 and
        2003, 124*f*
    remote sensing data, 120
    report, 116*f*
    scars, 117*f*
    Seattle–Tacoma, 118, 119*f*
    in Seattle urban area, 122–126
Land surface temperature (LST), 170
    calculation from remote sensing,
        171–174
    standard deviation of, 186
    thermal infrared, 171
Land use and land cover (LULC),
        125, 176
Las Vegas, orthoimage in, 15*f*

Las Vegas Valley
    Landsat scenes, 42*f*
    percent impervious surface
        2002, 44*f*
    temperature in, 177*f*
    urban impervious surface, 148*f*
Lead (Pb), 91, 93*f*
Lightning-triggered fires, 156
Linear split-window algorithm, 174
Low Earth orbit instruments, 132

M

Maximum peak-height (MPH), 86
    algorithm, 87
Measurements of pollution in the
        troposphere (MOPITT)
        instrument, 132–133
Medium-density built-up land, 124
Medium-resolution imaging
        spectrometer (MERIS), 86
MetOp-A satellite, 137
Microcystis cyano-dominant water, 87
Military intelligence satellite, 8
Minimum noise fraction (MNF), 32
    components, 33*f*
Mixing layer, 169
Model file, 41
Moderate-resolution imaging
        spectroradiometer
        (MODIS), 154
    data, 67
    geographical collocation of, 160
    instrument, 132
        spatial resolutions of, 136
    map global urban
        accuracy assessment for global-
            scale product, 70–71
        classification of global urban land
            cover, 68–69
        classification tool, 66–67
        MODIS data, 67
        training data, 67–68
        urban ecoregion, 68
        urban extent definition, 66
    MOD500 and MOD1K, 61
    sensor, 174
Moderate-resolution sensors, 25, 28
MODIS AOD and PM$_{2.5}$, 160–163

MODIS 500 m, continental views of
        Asia, 71*f*
    North America, 70*f*
MODIS TSS measurements, 98–100
Multiangle imaging spectroradiometer
        (MISR) instrument, 132
    retrieval of, 136
Multilayer perceptron (MLP), 46
Multilayer perceptron neural networks
        (MLPNN), 46
Multiresolution land characteristics
        consortium (MRLC), 52–53
Multitemporal remotely sensed
        data, 55

N

Name file, 38, 41
National Geophysical Data Center
        (NGDC), 64
National Land Cover Database (NLCD)
        characterization, 52–59
    changes in the United States between
        2001–2011, 57–59
    development of product in the
        United States, 52–54
    updated impervious surface product,
        54–57
National Ocean and Atmospheric
        Administration (NOAA), 55,
        63, 112, 133
    estimate flood effects, 113
    satellites, 171
Natural hazards
    debris flows, 105
    and disasters, 105–106
    dust storms, 155–156
    effects of, 115
    floods, 105
    hurricanes, 107
    landslides, 105, 119
    rise in sea levels, 107
    tsunamis, 107
    in urban area, 106
    use of remote sensing, 106
    wildfires, 107
Natural Terrain Landslide Inventory,
        117
Near-infrared (NIR), 79

Neural network architectures, types
 of, 46
Neural-network structure, for
 classification, 47*f*
Nighttime light image, 56–57, 64
Nitrogen oxides (NO_x), 143–146
 aircraft measurements of, 146
 algorithms to measure, 144
 concentrations of, 145
NLCD 2001, 53
 classification and regression trees
 (CART), 54
 mapping regions, 53
 urban land cover, 53
NOAA-AVHRR images, 112
 estimate flood effects in
 Bangladesh, 113
 flood depth estimate, 127
Non-point sources, 77
 polynomial regressions of, 94*f*
Nonurban land, loss of, 7
Normalized difference vegetation index
 (NDVI), 12, 43, 120, 148, 176
 in Las Vegas Valley, 148*f*
 spatial variation of, 184
 vegetation density, 184

O

Object-based image analysis, 16
 flowchart of, 17*f*
Oil and grease, 91, 92*f*
Open space land cover, 13
Operational land imager (OLI), 26
Operational linescan system (OLS), 63
Organic acids, 153
Orthoimages, locations of seven
 high-resolution, 39*f*
Ozone (O_3), 129–130, 146–147
 air quality observations of, 147
 daily maximum surface, 149*f*
 emissions of, 132
 formation of, 130
 reduce, 144
 tropospheric, 146
Ozone monitoring instrument
 (OMI), 132
 NO_2 algorithm, 144
 in ultraviolet, 146

P

Particulate matter (PM), 129, 153
Phytoplankton pigments, 81
Planck's equation of radiation, 171
$PM_{2.5}$ distribution, assessment of,
 141–143
Point sources, 77
Pollutant, intercontinental transport of,
 163–167
Pollutant Loading and Removal Model
 (PLRM), 91
Polynomial regressions of non–point
 source, 94*f*
Principal component analysis
 (PCA), 18
 rules, 19

Q

Qualitative tracking, 157
QuickBird (satellite), 8, 106, 114
 imagery, 12–14
 NDVI imagery, 12
 satellite images acquired from, 25*f*
 spatial resolution of, 9

R

Radiant surface temperature,
 173–174
Recurrent networks, 46
Regression tree algorithm,
 34–45, 37*f*
 advantages of, 36
 example of, 40*f*
Regression tree model, 54–56
Relative humidity (RH), 157
Remote sensing data, 4, 8
 advantage of, 9
 characteristics of, 9
 use of, 9
Remote sensing technology,
 79, 170
Retrieval, 132
Root mean square error (RMSE), 31
Rs spectra of Frisian water, 83*f*
Rural areas population, 1
 and urban population, 2*f*

S

Sahara desert, aerosols, 155–156
Satellite instruments
    characteristics of, 134*t*–135*t*
    spectral range of, 80*f*
Satellite measurement, 156
Satellite remote sensing, 59, 154
    medium-resolution, 133–138
    monitor aerosol cross-continental
        transport, 156
    systems, 24*t*
    used for, 154
Satellites
    earth observation, 8
    Envisat, 136
    ERS-2, 133
    first-generation sensors, 8
    for hazard mapping and
        monitoring, 106
    high-resolution images from, 9
    IKONOS, 8, 106
    images, 25*f*
    land remote sensing, 8
    MetOp-A, 137
    military intelligence, 8
    QuickBird, 8, 106, 114
    retrievals, 132–133
    sensors, 81*t*
    SPOT, 116, 118
    Terra, 133, 136
    uses of, 157
    WorldView-2, 114
Satellite systems
    current high-spatial-resolution, 10*t*
    for global air quality assessment,
        156–157
Scanning Imaging Absorption
        spectroMeter for Atmospheric
        CHartogaphY (SCIAMACHY),
        132, 136–137
Seattle area
    estimate impervious surface, 39*f*
    ISA estimate, 45
    Landsat scenes for, 42*f*
    percent impervious surface in, 44*f*
Seattle–Tacoma (United States), 118, 119*f*
Self-organizing map (SOM), 46
Shortwave infrared (SWIR), 97

Solar backscatter, 131
Space-based remote sensing, 8
Spatial distribution patterns, 59, 76
Species observations, 138–141
Spectral mixture analysis (SMA), 29
    goal of, 30
    urban landscape, 30*f*
SPOT (satellite), 116, 118
    change images, 117*f*
    high-resolution-visible (HRV), 8
Storm water management, 11
Supervised method, 53
Support vector machine (SVM), 23,
        48–49
    basic idea of, 48
    three important properties of, 49
Surface layer, 169
Swimming pools, spectral features, 19
Systematic error (SE), 45

T

Taklimakan Desert, dust event, 154
Tampa Bay
    DOQQ in, 15*f*
    impervious surface area (ISA)
        estimation, 41
    overview, 90
    watershed, 42*f*, 44*f*, 90
        temperature in, 177*f*
TerraSAR-X data, 73–75
Terra satellites, 133, 136, 154
    near-polar orbits, 174
Terrestrial ecosystems, 52, 54
Test file, 38
Thematic Mapper (TM), 173
Thermal infrared (TIR), 171
    emission, 131
    TES in, 146
Thermal infrared sensor (TIRS), 26
Total dissolved phosphorus (TDP), 91, 92*f*
Total Kjeldahl nitrogen (TKN), 91, 92*f*
Total nitrites and nitrates ($NO_3 + NO_2$),
        91, 92*f*
Total nitrogen (TN), 91, 92*f*
    spatial distribution patterns of, 95
Total ozone mapping spectrometer
        (TOMS), 131
    advantage, 147

Total suspended solids (TSS), 81, 91, 93*f*
    concentration estimate, 89–90
Trace gases, 138–140
Tropospheric emission spectrometer
        (TES), 132
    Fourier transform infrared emission
        spectrometer, 137
Tropospheric ozone, 146
Tsunamis, 107

U

Universal transverse mercator (UTM)
        projection, 12
Unsupervised method, 53
UN Wall Chart of Urban and Rural
        Areas, 1
Urban agglomerations, 51
Urban areas, 153
    agricultural water usages in, 78
    assessment of water quality, 77–103
    base on biomes, 182
    in DE biome, 184
    industrial water usages in, 78
    land cover in, 28
    land surface temperature (LST), 170
    moderate-resolution satellite data
        in, 26
    people move into, 1
    population in, 1, 3*f*
    and rural population, 2*f*
    surface temperature between rural
        and, 186
    surround by deserts and xeric
        shrublands, 184
    VIS components in, 36
Urban atmosphere, 169
Urban classification study, 17
Urban ecoregion, 68
Urban ecosystem, 68
    biophysical characteristics in, 3
Urban extent definition, 66
Urban features, 169
Urban heat island (UHI), 2, 169–190
    analysis of, 178–182
    changes of, 186
    climate impacts of, 188–189
    compare amplitude of, 182
    for DE, 184

effect. *See* urban heat island (UHI)
        effect
    in high-density urban areas, 176
    mean intensity of, 184
    quantification of, 174–187
    study for, 187
    urban vegetation on, 176
Urban heat island (UHI) effect
    across continental United States,
        182–185
    analysis of global, 185
    characteristics of, 170
    in Europe, 185–187
    global aspect of, 187–188
    local scale, 175–178
    spatial character of, 182
Urbanization
    effect of, 3
    issues, 7
Urban land classification, 46–49
    artificial neural network (ANN),
        46–48
    support vector machine, 48–49
Urban land cover classification, 15*f*
    decision tree model format, 14–15
    mapping, 12–14
    object-based image analysis for,
        16–22
        flowchart of, 17*f*
    overall procedures flowchart for, 21*f*
Urban landscape, 169
    characteristics of, 7–22, 23–28
    characterization of, 29–45
    fundamental components of, 29*f*
    general approach of SMA for, 30*f*
    high-resolution image, 38*f*
    mixing features of, 28–29
    role of remote sensing, 7–9
    tool to characterize, 34
Urban polygons, 183
Urban population, 51
    in millions, 3*f*
    water quality issues, 77–78
Urban–rural temperature differences,
        181*f*, 181*t*
Urban seeds, 74
Urban sprawl, 51
Urban structures, 169
    analysis of, 12

U.S. Environmental Protection Agency
  (EPA), 161
U.S. Geological Survey (USGS), 14, 53, 65
U.S. Geological Survey National Land
  Cover Database (NLCD),
  178–179

V

Vector analysis, 55
Vector Map Level Zero (VMAP0),
  61, 62t
Vegetation density, 184
Vegetation-impervious surface-soil
  (VIS), 28, 29f, 34
  conceptual model, 66
Volatile organic compounds (VOCs), 130
  photochemical oxidation of, 146

W

Water, backscattering characteristics
  of, 79
Water conditions in Dai Lake, 78f
Water quality
  analyze, 80

assessment for china's inland lake
  Taihu, 100–102
detection of, 81t, 84t
deterioration, 77
empirical and semi-analytical, 80
factors affect, 79
in Lake Taihu, 100
map, 95–102
parameters of inland, 80–85
parameters used to measure, 83
problems, 2
urban population usage, 77
Water resources
  degradation of, 77
  non–point sources, 77
  point sources, 77
World population between 1950 and
  2050, variations of, 2f

Y

Yellow substances, 88

Z

Zinc (Zn), 91, 93f